Andrews
Thermodynamics

DATE DUE			
FEB 1 73			
JUL 20 '73			
FEB 13 74			
MAR 4 74			
MAR 21 74			
4-17-87			

THERMODYNAMICS: PRINCIPLES AND APPLICATIONS

THERMODYNAMICS: PRINCIPLES AND APPLICATIONS

FRANK C. ANDREWS

Crown College
University of California, Santa Cruz

WILEY—INTERSCIENCE

A DIVISION OF JOHN WILEY & SONS, INC.
NEW YORK · LONDON · SYDNEY · TORONTO

TO JEANIE

PREFACE

It is almost an act of intellectual arrogance to write another thermodynamics text these days, so many and varied are the texts already available. This abundance of books is due to several causes: the subject is so basic that all physical scientists and engineers and many biological scientists study it. The subject is difficult enough that a person often feels at home with it only after he has been exposed several times or perhaps not until he has taught the subject. The beauty of thermodynamics, however—its innumerable, varied results all derived from so little—attracts a variety of minds, from the most practical experimentalist to the most abstract theoretician. So it is natural for a host of teachers to want to convey their understanding and enthusiasm to others by writing books.

This book arose not only for these reasons but also from concern over the amount of time being spent in the usual pedagogy of thermodynamics. In a day when the molecular structure of matter occupies an ever greater number of scientists and engineers one just cannot afford to luxuriate in one course after another on thermodynamics. Too often watered down thermodynamics is fed to students at several levels, starting in the freshman year. It may be only in graduate school that they unlearn the half-truths of the earlier "ideal gas" presentations, but half-truths are hard to unlearn, and the series of learning and unlearning steps is frustrating and time consuming.

This book represents a first treatment of thermodynamics that follows the usual introductory courses in chemistry, calculus, and physics. Its spirit is to cover the fundamentals correctly so that what is learned here need never be relearned but can be built on confidently in any future work. The first part is a treatment of the principles of thermodynamics divorced from their applications. Little attempt has been made to do a few first law problems after introducing the first law, then to do a few second law problems after introducing the second law, etc. That would be the usual approach, and it has the virtue of ever-present variety and motivation. However, it has the serious fault of ignoring the end-purpose of the course, namely to enable the student to use the *entire machinery* of thermodynamics to organize, interpret, and predict an enormous variety of properties of matter. Only after the complete machinery of thermodynamics is presented does this book move on to its second part, a study of selected properties of matter as organized and interpreted by thermodynamic principles. The properties chosen are basic to the disciplines of chemistry, biology, engineering, physics, and earth sciences. Here one is free to concentrate on solving problems and interpreting data, using any or all of the tools of thermodynamics as dictated by convenience.

The first part of the book may seem like heavy, dry going; certainly it will not all be absorbed at once. It will take on real meaning only when it is read over and over during the study of the second part. It is hoped that this organization will avoid a common fault of many science courses: the lecturer is so fascinated by the basic theory that he spends most of his lectures deriving and discussing fundamental equations and their implications. At the same time the students are supposed to be understanding this material. They are working problems to show how the theory is used in interpreting important physical phenomena. Both the theory and its use are novel and difficult. The student's plight is made worse by the emphasis on theory in the lectures and on practical applications in examinations and in his subsequent career. Too often students decide simply to accept the basic theory because of the authority of the text or the lecturer and to view problem solving as possible only when some formula has been learned into which the data may be plugged. That viewpoint is bound to fail. There are *so many* ways that thermodynamics can be used with *so many* phenomena that involve *any real system* found in nature ! The only hope is to learn how to think through new problems as they arise with a confidence based on understanding the entire structure of thermodynamics. This is not easy, but at least it is a goal worthy of the effort.

To this end it might be found helpful to hurry through Part I as fast as possible in order to get to the applications in Part II. There is surprisingly little one needs to *know* from Part I in order to begin Part II—just the material of Section 16. All the rest is derivations and discussions. In fact, it might prove better for some readers to start the book with Section 16 and to use the first 15 sections simply as reference material for Part II.

This is a book on the fundamentals of macroscopic thermodynamics. I consider it to be suitable for the usual first undergraduate course in thermodynamics for science or engineering. If it were followed by a second short volume that treated the applications more specialized to a particular discipline, the result would be a useful senior or first-year graduate thermodynamics course. On the other hand, a combination of this book and my short (and highly recommended) text *Equilibrium Statistical Mechanics* (John Wiley & Sons, New York, 1963) should be useful for a thermodynamics and statistical mechanics course for seniors or graduate students. These two books are designed to complement each other. I am convinced that significantly clearer thinking results when macroscopic thermodynamics is studied separately from statistical mechanics, which is based on the microscopic picture of the molecular structure of matter. When these two approaches are blended, the result too often is a blur.

Thousands of people have come to love thermodynamics, to be tantalized and fascinated by it. Thousands more have come to hate it. I am in the first group, and if the reader can join me there this book is a success.

I want to thank the many students who struggled through early versions of this book. A number of scientists have reviewed the early drafts, and I especially thank Joseph E. Mayer and Peter F. Linde for valuable and creative comments on the manuscript. I also sincerely thank my father for exciting my interest in and love for thermodynamics.

FRANK C. ANDREWS

Santa Cruz, California
January 1971

CONTENTS

PART I. PRINCIPLES OF THERMODYNAMICS

PART II. APPLICATIONS OF THERMODYNAMICS

APPENDIXES

INDEX

THERMODYNAMICS: PRINCIPLES AND APPLICATIONS

PART I

PRINCIPLES OF THERMODYNAMICS

WHAT IS THERMODYNAMICS?*

There are ways to change the properties of matter that are not accounted for by the classical physics of mechanics, electricity and magnetism, optics, etc. Simply putting a burner under a system can change its properties dramatically; yet the interaction between burner and system is inexplicable in terms of these classical disciplines. We call the burner system interaction a *heat* interaction, rather than a *work* interaction, and there is no place for heat in the classical disciplines. A billiard table covered with rolling balls when left alone can be studied briefly by Newton's mechanics, but what is the final state of the system of table and balls? When the balls have stopped rolling, where has their energy gone? We say this energy has become additional *thermal energy* of the table and balls through *friction*, but there is no place for either thermal energy or friction in the classical disciplines. An additional variable is needed to describe systems if processes involving heat and thermal energy are going to be treated. Such a variable is the *temperature*.

Thermodynamics, then, is a branch of physical science concerned with relating and predicting the various properties of macroscopic matter, especially as they are affected by temperature. All the observable macroscopic properties may be studied: the mechanical properties (shape and size of a body); the hydrodynamic properties (pressure and viscosity); the thermal properties (heat capacity and thermal conductivity); the electric and magnetic properties (susceptibilities and dielectric constants); the chemical properties (equilibrium constants and rate constants); and many others.

---

* "To me, thermodynamics is a maze of vague quantities, symbols with superscripes (sic), subscripes, bars, stars, circles, etc., relationships arising from the ideal gas, but getting changed along the way, and a dubious method of beginning with one equation and taking enough partial differentials until you end up with something new and supposedly useful (if that doesn't work you try graphing). I have the impression, however, that to other people, thermodynamics is a logical study of the effects of temperature and pressure on matter and on energy. Also, it must have more subtle aspects, such as entropy and is the universe running down, . . . Then there are some processes that haven't been studied completely and these are the most challenging of all." (From an advanced graduate student essay entitled *What is Thermodynamics.*)

3

The properties and processes studied in thermodynamics seem to fall naturally into two divisions: When the study is limited to the ultimate equilibrium condition reached by matter when it is left alone, it is called **equilibrium** or **classical thermodynamics**. Such properties as pressure, heat capacity, and chemical equilibrium constants are treated. **Nonequilibrium thermodynamics**, on the other hand, considers the properties which govern the *approach* to the equilibrium condition, such as viscosity, thermal conductivity, and rate constants for chemical reactions.

Thermodynamics is a "phenomenological" discipline in that it treats phenomena and not their causes. No attempt is made to seek a mechanistic world in which molecules underlay the observable properties. In a sense, this is an advantage, because the thermodynamics is relatively simple and is general; i.e., it is independent of the picture of molecules. Its "truth" rests on only the belief that carefully controlled physical experiments will yield the same results tomorrow that they did today. Because thermodynamics gives no intuitive feeling of causality, modern science students brought up to analyze all experiences in terms of their causes might feel uncomfortable with it. Almost everyone who contributed to the development of thermodynamics was profoundly interested in the molecular picture too. But, by definition, those theories which show how the macroscopic world arises from the microscopic fall into the discipline of *statistical mechanics*. If they are any "deeper" than thermodynamics, it is only because they "explain" one large part of our experience in terms of a useful and appealing picture. But, of course, no amount of "explaining" one set of phenomena in terms of another can push the question "Why?" into a corner. Knowledge of the regularities found in the macroscopic world is necessary to any scientist. Both thermodynamics and statistical mechanics deserve serious study.

The main reason a macroscopic science is useful is that we live in a macroscopic world. If humans were the size of viruses, "macroscopic" studies such as classical mechanics, electricity and magnetism, and thermodynamics might have very little utility, except to our astronomers and cosmologists. As it is, our macroscopic world is much simpler than the microscopic. Macroscopic measurements are averages of the behavior of astronomical numbers of molecules. This leads to the regularity and predictability of the bulk matter studied in thermodynamics. As Bridgman said (Bridgman, p. 3): "The laws of thermodynamics have a different feel from most of the other laws of the physicist. There is something more palpably verbal about them—they smell more of their human origin." Certainly, as humans we cannot complain about that!

Any modern presentation of thermodynamics includes one view of a finished product that has been crafted and polished for hundreds of years. It is the result of a continuing interplay of the inductive-deductive type for which science is famous. After making a necessarily finite number of observations of the behavior of matter, certain generalizations or postulates were made from this experience. By using mathematics and logic, the many consequences of these assumptions were found, and these, in turn, checked against experience. Where the consequences were found to disagree with nature, the postulates were modified or scrapped. After an enormous effort, started in earnest in the nineteenth century, this interplay of assumptions, mathematics and logic, and experiment has led to the modern, polished theory of thermodynamics, about which Einstein said:

A theory is the more impressive the greater the simplicity of its premises is, the more different kinds of things it relates, and the more extended is its area of applicability. Therefore, the deep impression which classical thermodynamics made upon me. It is the only physical theory of universal content concerning which I am convinced that, within the framework of the applicability of its basic concepts, it will never be overthrown.*

Finally, despite contrary opinions that students sometimes acquire, it must be emphasized that thermodynamics treats *real* systems made up of *real* matter.† There is nothing idealized about the subject thermodynamics studies; it is the " real " world of experience. Sometimes physical properties have to be approximated mathematically in the solution of particular problems. Of course, the results then only approximate what will be found in nature, but these limitations are inherent in all of science. One can go with thermodynamics a long, long way toward precision and order in viewing the physical world.

It is a pity that in a book like this one, the joy and excitement of discovery, of human beings groping to order their experience, is not present. Instead, the polished discipline is laid out one step at a time in such a way that the whole thing looks almost obvious. It was not obvious, however. Men thought and fought and wept and laughed, and even died in despair over the ideas that look so dull on these pages.

SUGGESTED READINGS

Bridgman, P. W., *The Nature of Thermodynamics*, Harper & Brothers, New York, 1961, pp. vii—13.

Dixon, J. R., *and* A. H. **Emery**, Jr., *American Scientist*, **53**: 428 (1965).

Lewis, G. N., *and* Merle **Randall**, *Thermodynamics*, (2d ed. revised by K. S. Pitzer and Leo Brewer), McGraw-Hill Book Company, Inc., New York, 1961, preface.

Pippard, A. B., *Elements of Classical Thermodynamics*, Cambridge University Press, London, 1957, chapter I.

* Quoted in *Albert Einstein: Philosopher-Scientist* (P. A. Schlipp, Ed.), vol. I, p. 33. Harper & Row, New York, 1959.

† Or at least as real as anything else, without getting ourselves into an epistemological quagmire.

2 SYSTEMS AND PROPERTIES

This chapter discusses the definitions of several common words, and in particular considers the meaning of a thermodynamic property of a system. Since careful definitions help in clear thinking, any student of science (or anything else, for that matter) who uses a word should be ready and able to define it. However, there need be no fetish made over concise, one-sentence definitions to be memorized for later parrotting on examinations. Instead, definitions may happily be a paragraph or two of relaxed description which give the flavor of a word or tell how a quantity could be measured. No framework of definitions should be considered sacred; words are practical tools for clear thinking. It is always possible to come up with cases which are hard to fit into an established set of definitions. Rather than trying to force a fit, one is wise either to refine the existing definitions or else to use new words in discussing the new cases. Arguing over definitions is not so sterile as is often supposed, so long as it is realized that this is indeed what is being done. The criterion of good and bad definitions is simply their pragmatic value in making for better communication, for clearer use of the language. Often people can agree on the reality of a situation and simply not agree on the particular words with which to label the situation. In that event, perhaps everything of importance has been done, and the labeling is unimportant.

In many thermodynamics problems an important first act is to decide what will be called the **system**, which is no more than that part of the physical world to which we direct our attention. Frequently it is the contents of a well-defined geometrical volume of macroscopic dimension. It may, however, be a given mass of fluid moving through a pipe or any other part of the world at which we especially want to look.

The **surroundings** are all the rest of the universe which can in any significant way affect or be affected by the system. Thus what one calls the surroundings may change from one problem to the next, even with the same system, as the word "significant" has meaning only within the context of a particular problem. An influence that might be of crucial importance in one phenomenon might be orders of magnitude too small to affect some other phenomenon significantly.

An *open system* is one which may exchange matter with its surroundings; a *closed system* may not.   Thus the cubic meter of air in a particular corner of a laboratory is an open system, since it is constantly mixing with air from the rest of the lab.   The air contained in a box, however, is a closed system if the box is impervious to air.   If, though, a wall of the box is a membrane through which air can pass, the system is open.   If the membrane passes only nitrogen and argon and not oxygen, then the system is open to the nitrogen and argon in the air, but closed to the oxygen.

In Sec. 1 reference was made to "the various properties of a system" as if it was obvious what was meant by the phrase.   In science, a *property* of a system is defined by specifying a procedure for performing a measurement the result of which is the value (usually numerical) of that property.   Of course, if the "property" is to prove very useful, it should have one or more of the following characteristics:

1. Either repeated measurements of the same property yield the same result, or else it is meaningful to talk about "the property changing its value."

2. There are several different procedures which yield the same resulting value, thus making it useful to say that some property of the system is actually being measured; that is, we are talking more about the system than about our own procedures of measurement.

3. The property is sufficiently related to our everyday experience that we feel at home talking about it; we develop an intuitive feeling for it.

These three characteristics are illustrated by our measurement of the "mass" of a billiard ball.   One way to measure this is with a set of weights on an analytical balance.   If this is done repeatedly with different weights and on different balances, the resulting values of the "mass" are almost the same.   Differences are attributed either to random variations in experimental conditions or to systematic differences in the sets of weights or the balances used.   It is also possible to determine the mass from the extension of a calibrated spring supporting the billiard ball in the gravitational field of the earth.   It may be found, too, from measuring the length of time the ball takes to fall a given distance from rest under the earth's gravitational attraction.   These other ways to "measure the same thing" yield the same numerical result, or else we never would have concluded they were measurements of "the same thing."

There is no a priori reason why our experience should show its remarkable consistency and reproducibility.   It is one of the most remarkable features of that experience.   This is the reason why it is so useful for men to create mental models of an external "reality" filled with relatively permanent objects having certain properties.   The ball "has a certain mass;" our experiments are simply means for us to learn approximately what that mass is.

Sometimes one sees a distinction made between ***fundamental properties*** of a system, which are directly or easily measured, and ***derived properties***, which are useful but are obtained from fundamental ones by some kind of mathematical functional relationship.   However, there is an arbitrariness to this distinction.   The position of a particle is almost certainly a fundamental property; most physicists would probably call the velocity of the particle a fundamental property too, even though its measurement involves measuring several positions and at least one time interval.   On the other hand, the kinetic energy, calculated from the mass and the square of the velocity, is likely to be called a derived property.

Most thermodynamic properties are usefully categorized as either extensive or intensive.   ***Extensive properties*** are those whose measurement requires looking at the system as a whole; ***intensive properties*** are well defined in each small region of the system.   Thus volume and mass are extensive properties, since they cannot be determined without examining the system as a whole.   When one says, however, that intensive properties are defined in each small volume, the word "small" means small with respect to the total volume V.   In a macroscopic discipline such as thermodynamics, any "small" region considered must still be large compared to the volume available to a molecule; that is, it must still be a macroscopic region and these small regions must still in a sense be large.

The idea of intensive, macroscopic properties therefore requires some amplification.   A good example of an intensive property in a gas is the density.   Density is defined by an experiment which determines the total mass contained in a given volume.   The volume may be small compared to the total system volume V, but it must be large enough to contain an enormous number of particles.   If it were not, if for example the volume chosen were a *cubic angstrom*, there would be either one particle or no particle in the volume, and the "density" would fluctuate rapidly.   A density so defined would not be a thermodynamic property.   There might be times when such a quantity was what one needed to know for a particular purpose, but its use would involve the loss of the great simplicity that thermodynamics affords.   Instead of the nice, stable, macroscopic density found by taking, say, a *cubic millimeter* for the volume, the complicatedly fluctuating property would be most difficult to predict.   Volumes chosen in density measurements might thus be small compared to V, but must be large compared to V/N, where N is the total number of particles in the system.*

* The expected number of molecules contained in a subvolume v in a large volume V is of course Nv/V, where N is the total number of particles in V.   The number actually found in v is expected to fluctuate randomly about Nv/V, the difference or fluctuation being of the order of the square root of the number Nv/V expected.   Thus in a subvolume of a gas in which 100 molecules are expected, the fluctuation would be around 10 and the number

One might ask, then, how the density can be defined at each *point* in the system, as is often done. The answer is that the definition is meaningful through a limiting procedure: Smaller and smaller volumes, always of course much larger than V/N, centering on the point of interest are chosen. If the value of the ratio of mass to volume levels out more or less for small volumes before the fluctuations become noticeable, that plateau value is said to be the density at the point. Defining intensive variables at points in a system is always based on such a procedure.

An example will show what was back of the careful choice of wording of the preceding paragraph. We are all familiar with the idea that a gas in a field-free container assumes a "uniform density." Of course, this uniformity is found only if regions much larger than V/N are examined. However, wherever in the vessel such a volume is considered, the same density will be observed. This is not so if the gas is in a gravitational field; there the density changes with position. To define the density at a particular point, one must consider a volume much larger than V/N, yet one whose dimension in the direction of changing density is much less than the distance over which the density changes appreciably due to the field. This is not a stringent requirement, since air at room temperature and 1 atm of pressure contains some 24,000,000 molecules in a cube only 0.001 mm on a side. In a very severe shock wave, such as that generated by a supersonic airplane, it might be possible to find an appreciable density variation over such a small distance.

It is clear that even with the 1-Å cube, a nonfluctuating value called density could be obtained by averaging the number of particles found in the volume v over a period of time τ long compared to the duration of a fluctuation:

$$\rho = \frac{1}{\tau} \int_0^\tau \frac{n(t)}{v} \, dt,$$

where $n(t)$ is the number of particles found in v at time t. This number fluctuates constantly as particles randomly enter and leave v. Making v larger decreases the frequency of significant fluctuations, and making τ longer smears out their effect. Time averaging complements volume averaging—longer times permit smaller volumes, and vice versa. Of course, if the density is actually changing with time, perhaps because one is looking inside an evacuated chamber which has recently been opened to the air, then one must be careful if he is employing time averaging. The averaging time τ

---

actually found in the subvolume would probably lie between 90 and 110. In a subvolume in which 25,000,000 molecules are expected, the likely fluctuation would be only around 5000, a number which would probably be undetectable. The subject of fluctuations in statistical mechanics is developed in Andrews, *Equilibrium Statistical Mechanics*, Sec. 33, John Wiley & Sons, New York, 1963.

must be short compared to the time in which the density changes significantly; otherwise the choice of τ would affect the result. This is similar to the problem of air in a gravitational field discussed previously, only it is applied to time averaging rather than to space averaging.

Another example of an intensive property is the pressure p in a gas. This is defined by an experiment which measures the force exerted on a test surface. If the area of the surface is small, then its random bombardment by the gas molecules will give a very rapidly fluctuating force. Averaging the force over a period of time wipes out the effect of the fluctuations. If the area is large, the fluctuations are less significant and occur for shorter time periods. Again, there is an inverse connection between the size of the test surface and the duration of the measurement needed to achieve a macroscopic result. Again, if the thermodynamic pressure is itself changing, the measurement time must be short compared to the time of significant change in p. The pressure measurement involves a finite region of the gas, and defining the pressure at a point must be done through an extrapolation procedure much like that used for the density.

In summary, the fact that thermodynamic properties are macroscopic means the following:

1. They pertain to regions containing huge numbers of particles (atoms and molecules, or perhaps photons).

2. The time taken for a measurement is usually long compared to the times within which the measured quantity is fluctuating due to the random motion of the particles.

3. Energies involved are enormous compared to individual quanta.

With a few exceptions (such as temperature T) it is conventional to use capital letters to represent the values of extensive properties and lowercase letters to represent intensive properties (Lowercase t conventionally means time, or temperature in °C). Unless clearly noted, all symbols used in this book refer to the system and never to part of the surroundings.

It will prove necessary to work with properties (such as entropy) the definitions of which in terms of easily visualized fundamental properties are so abstract and unphysical that at first no physical insight whatever will be experienced. The only way to develop physical feeling for the formal theory presented in Part I of this book is to see how it is used to organize and relate the properties and behavior of matter. This is done in Part II. Thus one need not worry if Part I seems abstract and hard to remember. If it is returned to and reread over and over during the study and problem solving of Part II, it will slowly come to life. The physical interpretation of the formal structure of thermodynamics comes from its application to specific problems.

PROBLEMS

2-1. Prove the assertion that air at room temperature and 1 atm of pressure contains some 24,000,000 molecules in a cube 0.001 mm on one side.

2-2. A cube of what length would be expected to contain 100 molecules of the air of Problem 2-1?

3   THE STATE OF THE SYSTEM

The **state** of a thermodynamic system is specified by the values of all the properties of that system. Of course, there are as many properties as one can imagine measurements performable on the system. However, we say that we *know the state* of the system when we know the values of a set of properties such that: (1) Every time our system with these properties is subjected to a given environment (either stationary or changing), every feature of its subsequent behavior is identical.* (2) This is true for every possible choice of environment (stationary or changing).

The task in thermodynamics is then to find a set of the *minimum number* of properties in terms of which all the others can be computed, and then to find the rules for doing the computing. There is no way to tell on theoretical or philosophical grounds how many properties will comprise this minimum number or which ones they might be. One must look at the physical world to find out. For example, in the study of classical mechanical phenomena it was

---

* This statement refers only to *deterministic* scientific theories, such as classical mechanics, electromagnetism, and thermodynamics, which describe most simple macroscopic systems to a fantastic degree of accuracy. If a minimum number of properties are known at a particular time, then *all the others* can, in principle, be calculated both for that time and at all later times. However, when one looks at small systems, such as the atoms, molecules, and subatomic particles which make up macroscopic matter, the quantum theories which must be used for description are not deterministic. Instead, only *probabilities* of various values for the properties of these small systems can be computed from knowing their states. Matter in the small is at best probabilistic, while nonliving macroscopic matter commonly seems deterministic, and one of the most fascinating aspects of the study of statistical mechanics is to see how the determinism of bulk matter arises from the indeterminism of its enormous number of constituent particles. For microscopic systems, the phrase "every feature of its behavior is identical" would be replaced by "enough information is available to permit in principle the *best possible* prediction of its subsequent behavior."

found that knowing just the coordinates q_i was not enough to determine the various properties of a system, but if, in addition, the various velocities $\dot{q}_i$ were known, then the system's state would be defined.   All properties of the system, not only at the time the q_i's and $\dot{q}_i$'s were known, but also for all times to come, were in principle calculable from just this much information (and the conditions at the boundaries of the system).   Newton's great contribution was in part to establish this fact and to give the rules for performing the calculation, that is, the functional relations relating all properties of a classical mechanical system to the values of the q_i's and $\dot{q}_i$'s at a single time.   Experience therefore tells us that the state of a classical mechanical system is completely specified by the set of coordinates and velocities.   This specification is, of course, not unique; replacing velocities by momenta, for example, leads to an equally adequate description.

In thermodynamics the procedure is completely analogous; one must turn to experiment to find out what a minimum set of variables is; and the problem of specifying states is more difficult than for classical mechanics.   If we consider systems of ever increasing complexity, we soon reach some for which this problem has not been solved, even in principle.   Only in specially simple situations has it been solved, though, luckily, a careful consideration of these situations enables us to make some important and valuable generalizations to any situation.   These contribute to our physical intuition, which aids greatly in our understanding of any problem.

For example, the simplest system is one whose properties do not change with time, that is, one which is in equilibrium.*   The minimum number of variables needed to specify the state of an equilibrium system will be seen to depend on the kinds of phenomena being studied.   Examples of typical variables are temperature, volume, and number of moles contained in the system; but the specification is not unique.   Volume might be replaced by pressure, for example.   Equilibrium states are the easiest to specify and require the least information about the system.

Systems whose properties change slowly with time due to gradients in temperature, density, or some other quantity pose somewhat more complicated problems.   Here again, the number of variables in a minimum set will vary with the problem, but the number will be larger than in the case of equilibrium.   At the least, one would need to know the spatial dependence of temperature, density, or whatever property was nonuniform.

Beyond that, more and more complicated situations can be found for which the thermodynamic description of a state becomes complex to the point of impossibility.   One difficult case is a very dilute gas, such as the air at high

* The concepts of equilibrium and of temperature are discussed in considerable detail in Secs. 8 and 9 respectively.   Here their intuitive meanings are good enough for following the discussion.

altitudes, affecting, for instance, the trajectory and reentry of missiles and earth satellites. Another is the turbulent flow of liquid in a pipe or of air over an airplane wing. In fact, it is so difficult to characterize the state of systems in turbulent flow that thermodynamics is of very little use in studying such flow. However, these difficult problems are obviously easier to face if one starts with an understanding of conventional thermodynamics.

It is important to make clear that the properties of a system that we are talking about depend upon the state of the system at the time in question and not on something that happened to it in the past. True, the past history of a system is a description of how it got in its present state, but thermodynamics treats states only, and past histories are not part of the states. The volume of a system is the same whether it was made that size 4000 years ago and stayed there ever since or whether it has just now reached that size on being squeezed down from something much larger. This definition of thermo-dynamic properties in terms of measurements on the system at the present time and not over its past history is emphasized by a commonly used expression: Thermodynamic properties are called **functions of state**. They may be obtained from the state of the system now; they do not require knowing what the state was in the past.

Consider the volume of a system, denoted by V. Any infinitesimal change in the volume is denoted by dV, and dV is the differential of a function of state. This means that the sum (integral) of the infinitesimal volume changes *for any process* which starts in state 1 with the volume V_1 and ends in state 2 with the volume V_2 must lead to a total volume change which depends only on the initial and final states:

$$\text{Volume change} = \int_1^2 dV = V_2 - V_1. \tag{3-1}$$

The value of the integral (the right-hand side of Eq. 3-1) is absolutely in-dependent of the particular path or route or way in which the volume was changed; it depends only on states 1 and 2. It could have been changed slowly or quickly, directly or through some complicated oscillation. The differential of a function of state, which can be integrated like dV in Eq. 3-1 in the fashion commonly taught in calculus courses, is often called a **perfect differential**, a **total differential**, or an **exact differential**.

Differences in the values of state functions, such as $V_2 - V_1$, appear often in thermodynamics, and for brevity they are commonly written

$$V_2 - V_1 \equiv \Delta V. \tag{3-2}$$

The capital delta notation is best reserved only for differences in values of state functions and not for any other purpose.

It is obvious from the definition of state functions that there is zero total change in any function of state for any process which ends in the same state it started, in which case $V_2 - V_1 = V_1 - V_1 = 0$. This is sometimes stressed by writing the equation

$$\oint dX = 0 \tag{3-3}$$

where X is any function of state, and the $\oint$ means an integral around any closed path, that is, one which comes back to the initial state. A process which comes back to the original state is called a **cycle**.

A final confusion in language should be mentioned. It is true that once a particular group of classical particles has been specified as a system, the state of that system is given by the set of coordinates and velocities for the particles. However, in order to solve the equations of motion, the characteristics of the particles that comprise the system must be known. These characteristics are the masses, the moments of inertia, and the laws of force (for example, the particles, shapes and sizes) between them. These characteristics distinguish the particular group of particles in question from other groups, and they are also sometimes called "properties" of the particles. Something similar occurs in thermodynamics, naturally enough. Whereas temperature, volume, and number of moles are a minimum set of properties which may fix the state of a simple equilibrium system, the calculation of other properties of the system depends, of course, on what the system consists of—whether it is water or air, for example. The different values for the "properties" of water and air, such as heat capacities, molecular weights, vapor pressures, etc., enter the theory to give different results for different substances. It might even be wise to use different words for the unique properties which establish the material being discussed and for the quantities such as volume, temperature, and mole numbers which are studied experimentally. These latter could be called "variables" instead of "properties," though the distinction would be arbitrary.

4 WALLS

One refers to the **walls** of a system as that part of its surroundings which (a) immediately surround the system and (b) "through which" it is convenient to think of the changes of the system being imposed. There are various kinds of walls, just as there is a variety of ways a system may interact with its surroundings.

Most walls will, to some extent, "convey heat." If pancake batter is poured on a hot griddle, one of the "walls" separating the batter from its surroundings is the surface (in this case of essentially zero thickness) between the batter and the griddle. "Through this surface" an interaction occurs which dramatically changes the state of the batter. The interaction is not a mechanical one; the griddle does no work on the batter; it is a *heat* inter-action.* Heat interactions, the "flow" or "transmission" of heat from "hotter" to "colder" bodies, are commonly seen by us all. It is more or less easy to cause a heat interaction, depending on the nature of the walls. For example, it is easy to cause heat interactions for the pancake or for an object contained, say, in thin metal walls; a few seconds in an oven or immersed in an ice water bath changes their properties markedly. However, one can imagine making walls through which it is more and more difficult for the system to interact with the surroundings in this way, for example, by making more and more elaborate vacuum bottles with many insulating layers, evacuated spaces, and shiny surfaces. One can extrapolate this process to so-called **adiabatic** walls, that is, walls through which there are no nonmechanical (heat) interactions with the surroundings. Since we dislike idealized definitions, however, we call walls effectively adiabatic when the nonmechanical (heat) interaction they permit with the surroundings during the time of the experiment is negligible. The walls of space ships are sufficiently adiabatic so that the astronauts do not freeze in the very cold environment of space and do not roast during the fiery procedure of reentry. *An adiabatically contained system can be disturbed only by mechanical means (and not by heat).*

---

* A quantitative definition of heat will be given in Sec. 7.

Walls which are not adiabatic are called **diathermal** walls.   Two subsystems separated by a diathermal wall are said to be in **thermal contact**.   They may have a heat interaction through the diathermal wall.

There are a lot of different kinds of mechanical interactions between system and surroundings which can occur through adiabatic walls.   Material can be added to or removed from the system either physically or through permeable or semipermeable walls.*   Other mechanical means of interacting from the surroundings with a system "through or by means of the walls" are changing the size of the volume that holds the system, putting the system inside a coil and changing the magnetic field on it, putting the system between two electrodes and changing the electric field on it, putting the system in a centrifuge to change its gravitational field, breaking the bulk of the system up into smaller pieces and thus increasing its surface area, pulling on the system and thus changing its length, freeing a piston inside the system so that is can move and change the relative volumes of two subsystems, opening a stopcock in the system so that matter can move between two subvolumes, etc.

A system is said to be **isolated** if it is both adiabatically contained and has no mechanical action taken on it.†

We note that, if a system interacts strongly *with the walls* themselves, then thermodynamic analysis becomes difficult at best.   This is why problems involving sulfuric acid contained in iron vessels are not given in elementary thermodynamics books.   In this case, the vessels would most likely be best considered as part of the system.

---

* We shall consistently ignore this possibility and treat only closed systems unless we explicitly call attention to the fact that we are doing otherwise.
† The very concept of adiabatic and isolated systems has been well criticized by Teilhard de Chardin (*The Phenomenon of Man*, Harper & Brothers, New York, 1959, book I, chapter 1, part 2):

> "Looking at matter as such, that is to say according to its qualities and in any given volume—as though it were permissible for us to break off a fragment and study this sample apart from the rest . . . this procedure is merely an intellectual dodge. . . .   It is impossible to cut into this network, to isolate a portion without it becoming frayed and unravelled at all its edges. All around us, as far as the eye can see, the universe holds together, and only one way of considering it is really possible, that is, to take it as a whole, in one piece.

Teilhard's points are surely true on an absolute basis, but from the practical point of view, many systems can be treated as isolated with results completely in accord with experiment.   We take the strongly pragmatic view which has characterized the advance of science, which in this case permits breaking the universe up into somewhat digestible chunks.   If this were not possible, and if man had to take the universe as a whole, in one piece, he would certainly have given up the whole idea of science long ago as completely incomprehensible.

5 WORK

The amount of work we have to do and the amount of work we can get machines (or some other person!) to do for us have always been of great interest to mankind. Thermodynamics is a work-oriented subject; its origin was based in the study of heat engines, and it has retained a strongly practical flavor to this day. The definition and treatment of work as a precise quantity are familiar from studies of physics, but because of their great importance to thermodynamics, they deserve a careful review.

What we call work involves a great variety of processes. Each form of work has the feature of being completely convertible (in principle, but unfortunately not in practice) into any of the other forms. For example, gravitational work may be done on some water falling over a dam, thus accelerating it to a high speed. The water may then perform mechanical work by exerting a force through a distance against the turbine of an electric generator. The turning of the generator yields electric current which may be used to do electrical work, even at places far away from the source of the falling water. This electrical work could, for example, be converted into mechanical work through an electric motor, and this mechanical work could, in turn, lift some water up against the force of gravity. The whole example is sketched in Fig. 5-1. Each conversion of work from one form to another can be made nearly 100% efficient. In our example the water at the end could be lifted back up almost as high as it fell to begin the sequence. Of course, in reality there will be frictional losses from the motions of the water, the generator, and the engine, so the theoretical limit of 100% efficiency can only be approached, not reached.

This convertibility of one form of work into another makes it possible to find the equivalences between the various kinds by comparing how far each one, say, raises a standard weight against the earth's gravitational field. Thus, suppose part of the surroundings does a certain amount of work dW on the system, as suggested by Fig. 5-2a. This changes the states of both system and surroundings. We can in principle determine dW by rigging a mechanical (that is, no heat, no friction, 100% efficient) device, such as the one of Fig. 5-1, which does work on the surroundings, restoring them to their

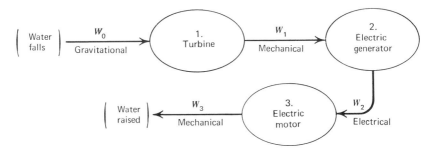

Fig. 5-1. Interconversion from one form of work to another.   In principle, each conversion can be made 100% efficient, so that at the end of the process the water, turbine, generator, and motor are in the same states in which they started.

original state, leaving as the sole external effect the change in height dh of a standard weight of mass M_0.   This is suggested by Fig. 5-2b.   The **amount of work** dW that was done on the system by the surroundings in the infinitesimal adiabatic process is defined to be

$$dW \equiv -M_0 g \, dh, \tag{5-1}$$

where g is the acceleration of gravity at the position of the weight.   The minus sign is needed because positive work done on the system leads to a lowering of the weight.   Using this approach, one can find experimentally the expressions for the work done in various processes, some of which are now discussed.

If the surroundings simply push parts of the system around, the work takes the form familiar from mechanics:

$$dW = \mathbf{f} \cdot \mathbf{ds} = |\mathbf{f}| \, |\mathbf{ds}| (\cos \theta). \tag{5-2}$$

Here dW is the work done on the system by surroundings which exert the force $\mathbf{f}$ against part of the system and cause the infinitesimal displacement $\mathbf{ds}$

Fig. 5-2. (a) Amount of work dW done on system, measured by (b) restoring surroundings to their original condition by doing work on them in such a way that the only effect in the surroundings is the change in height of a standard weight by an amount dh.

of that part.   The angle between the direction of the applied force and the direction of the displacement is θ.

An example of this kind of work arises if the system is a wire or other elastic object whose length l is changed to $l + dl$ owing to an external force f acting in the direction of the wire.   See Fig. 5-3.   The work dW done by the

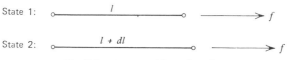

State 1: l

State 2: $l + dl$

Fig. 5-3. The stretching of a wire.

surroundings on the system is

$$dW = f\,dl, \tag{5-3}$$

since the force the surroundings exert is in the same direction as the displacement.

Another example of this kind of work arises when a three-dimensional system expands or contracts its volume under an external pressure.   For simplicity, let the system be a rectangular box of dimensions $h \times b \times l$ under an external pressure p_{ext}.   This p_{ext} is the force exerted by the surroundings on each unit of area of the system.   The total force exerted on the end of the system, whose area is $h \times b$, is therefore $p_{ext}\,hb$, and this is shown in Fig. 5-4a.

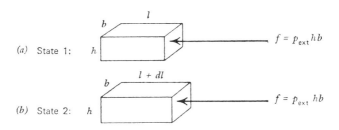

(a) State 1: h b l $f = p_{ext}\,hb$

(b) State 2: h b $l + dl$ $f = p_{ext}\,hb$

Fig. 5-4. The expansion of a volume.

Similar forces are exerted on the other faces of the system.   Suppose, now, the volume change is effected by the lengthening by dl of the system in the direction perpendicular to its end.   This is shown in Fig. 5-4b.   The use of Eq. 5-2 to calculate the work yields

$$dW = \mathbf{f} \cdot \mathbf{ds} = -p_{ext}\,hb\,dl. \tag{5-4}$$

The minus sign enters because the external pressure pushes back against the system, so the directions of force and displacement are opposite, and

$\cos \theta = -1$. A simplification in Eq. 5-4 may be made by noting that the infinitesimal change in volume dV for this expansion is

$$dV = hb(l + dl) - hbl = hb\, dl, \qquad (5\text{-}5)$$

so Eq. 5-4 may be written

$$\boxed{dW = -p_{\text{ext}}\, dV.} \qquad (5\text{-}6)$$

This same result, Eq. 5-6, is obtained for systems of arbitrary shape (see Prob. 5-5). This type of work is often called "pressure-volume work" or "p-V work," for short.

Yet another example of mechanical work is involved in overcoming friction. For example, let the system be the composite of a book and the table on which it rests. If we push the book over the table, we must exert a force in the direction of motion which overcomes the frictional resistance. The work done is given by integrating Eq. 5-2.

Another commonly encountered form of work is electrical. The difference in **electrical potential** φ is defined by the statement that work dW (as measured perhaps by a device like Fig. 5-2) is done by the surroundings in taking the infinitesimal charge dq through the potential difference φ:

$$dW = \varphi\, dq = \varphi i\, dt. \qquad (5\text{-}7)$$

where dq has been written as current i times duration of time dt. It is interesting that, depending on the definition of the system, one situation may give rise to electrical work on a system while a similar one leads to something else altogether (heat). In Fig. 5-5a, the system is defined as everything

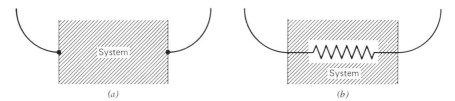

(a)   *(b)*

Fig. 5-5. (*a*) Electrical resistance is due to system.   (*b*) Resistance wire is part of surroundings.

between the two wires. Here the surroundings do electrical work $\varphi\, dq$ on the system, where φ is the potential difference between the junctions of the wires to the system. In Fig. 5-5b, electrical work is done on the resistance wire, which is considered part of the surroundings. With this choice of definitions, the interaction between system and surroundings is a flow of heat from the hot resistance wire to the system.

A similar expression is found for the infinitesimal element of gravitational work.   The **gravitational potential** ϕ is defined by the statement that work

$$dW = (\phi_2 - \phi_1)\, dM \tag{5-8}$$

is done by the surroundings in taking the infinitesimal mass dM from position 1 with potential ϕ_1 to position 2 with potential ϕ_2.

If there are interfaces within the system between liquids or solids and gases, or between liquids and solids, or between one liquid or solid and another, the interfacial or surface area can be increased adiabatically by doing work on the system.   For example, one may cut a solid into little pieces or break a liquid up into tiny drops.   The work done by the surroundings in creating infinitesimal surface area $d\mathscr{A}$ is found to be proportional to the increase:

$$dW = \gamma\, d\mathscr{A}, \tag{5-9}$$

where the proportionality coefficient γ is called the **surface tension** for the particular kind of interfacial surface being created.   Thus γ is simply the work required to create unit area of surface.

The work required to polarize a system either electrically or magnetically by increasing an external applied field is in general fairly complicated to determine.   For simple cases,* however, the element of electrical work done in polarizing a system takes the form

$$dW = \mathscr{E}\, dD, \tag{5-10}$$

where $\mathscr{E}$ is the electric field strength and D the electric displacement.   The similar expression for the work done in polarizing a system magnetically is

$$dW = \mathscr{H}\, d\mathscr{M}, \tag{5-11}$$

where $\mathscr{H}$ is the external magnetic field and $\mathscr{M}$ the magnetization.

If all these kinds of work can be done on a system, the element of work for an infinitesimal process is the sum of terms:

$$dW = f\, dl - p_{\text{ext}}\, dV + \varphi\, dq + (\phi_2 - \phi_1)\, dM + \gamma\, d\mathscr{A} + \mathscr{E}\, dD + \mathscr{H}\, d\mathscr{M}. \tag{5-12}$$

Most of the applications to thermodynamics treated in this book involve p-V work, so for brevity we frequently write Eq. 5-12 in the simpler form:

$$dW = -p_{\text{ext}}\, dV + X_i\, dx_i. \tag{5-13}$$

The last term in Eq. 5-13 represents the sum of all terms of the form of those in Eq. 5-12 which might be relevant to the problem at hand.

* Detailed discussions are given by H. B. Callen, *Thermodynamics*, John Wiley & Sons, New York, 1960, Chapter 14, and by G. N. Lewis, M. Randall, K. S. Pitzer, and L. Brewer, *Thermodynamics*, 2d ed., McGraw-Hill Book Company, New York, 1961, Chapter 31.

Once we have written Eq. 5-12, the method of Fig. 5-2 and Eq. 5-1 may be used to compare the magnitudes of the quantities appearing on the right-hand side. For example, if the force f is measured in dynes and the displacement l in cm, the $f\,dl$ term comes out to be in ergs. However, the external pressure might be measured in atmospheres and the volume in liters, so the $p\,dV$ term has units in liter-atmospheres. Similarly, if φ is in volts and dq in coulomb, the $\varphi\,dq$ term is in volt-coulombs or joules. The conversion of any of these work units into any other is accomplished through performing any experiment approaching 100% efficiency which converts the one kind of work into the other. The device of Fig. 5-2 is a simple method which converts all work contributions into gravitational potential energy. However, other conversion means, as illustrated in Fig. 5-1, for example, could be used just as well. The following equalities are found for various common units of work or energy: 1 joule $= 10^7$ ergs $= 2.778 \times 10^{-7}$ kilowatt-hour (kwhr) $= 0.009869$ liter-atmosphere (l-atm).

The symbol dW means an infinitesimal amount of work done on the system. The bar is placed through the d in dW because dW is not a perfect differential in thermodynamics. That is, the sum (integral) of dW over some path between states 1 and 2 does not take the form, Eq. 3-1, of a function of state 2 minus that same function of state 1. We prefer to use the symbol dX for a quantity X which is integrable in that sense. Some authors use the notation δW or DW, or simply W, instead of dW, and other authors do not bother to use different notation to distinguish perfect from imperfect differentials.

The work W done on a system is a measure of a particular *process*; the process, of course, takes a system from an initial to a final state, but the work describes the process and not the states as such. Everyone is familiar with such path-dependent quantities; for example, the *total distance* traveled in going from one place to another depends only in part on choice of places; it really describes the journey. It can assume *any value* larger than some minimum; only the minimum distance traveled is a function of the end points alone. Clearly, work is not a property of a system; work is meaningful only when referred to a process. Since it is not a function of the state of a system, it is meaningless to talk about so much work being "contained" or "stored" in a system. It is unwise even to say that work is "transferred" from one system to another, since the word connotes something tangible. *Work* refers to processes and not to systems.

The dW cannot be integrated as in Eq. 3-1. Instead, the sum of the infinitesimal contributions dW for a particular finite process in which the system changes from state 1 to state 2 is

$$W = \int_{1\text{ path}}^{2} dW. \tag{5-14}$$

Such a quantity is called a *line integral*, and its value depends upon the path taken between states 1 and 2.

That this is so may be seen by considering an example. For p-V work,

$$W = - \int_{1 \text{ path}}^{2} p_{\text{ext}}(V) \, dV, \tag{5-15}$$

where the fact that p is a function of V is noted explicitly. One might ask how it can be that the integral in this equation can have different values, when such definite integrals treated in calculus courses usually have unique values depending only on V_1 and V_2. The answer is that p_{ext} depends not only on V but also on one or more other variables (like temperature), and by controlling them during the process (through heat interactions which can heat up or cool down the system) one is free to make the integral assume any value.

For a specific example, in Fig. 5-6 we indicate state 1 (p_1 and V_1) and state 2

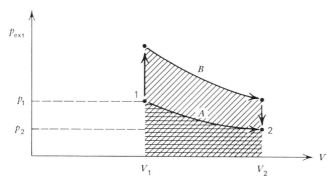

Fig. 5-6. Different paths between states 1 and 2 having different values of work.

(p_2 and V_2) on a plot of p_{ext} versus V. Suppose some simple process represented by the line A takes the system from state 1 to state 2. The work W, as shown by Eq. 5-6, is the negative of the area under line A (shown by the horizontal hatchmarks). Consider a different process, B, in which the system (a gas, say, contained in diathermal walls) in state 1 is first placed in a furnace for a heat interaction. In order to keep the volume fixed at V_1, p_{ext} must be increased as the gas gets hot. When p_{ext} gets sufficiently high, the system is removed from the furnace and the natural expansion is taken from V_1 to V_2. The system can then be allowed to sit around at fixed volume until it cools (that is, a heat interaction with cool surroundings) enough that pressure p_2 is reached. The area under curve B (shown by the diagonal hatchmarks) is clearly much greater than that under A. The two processes lead from the same initial state to the same final one, but involve different amounts of work.   In fact, one could imagine an infinite number of paths going from state 1 to state 2, involving every possible amount of work from minus to plus infinity (see Prob. 5-4)!

It is easy to see from Fig. 5-6 how the criterion, Eq. 3-3, for integrating state functions is disobeyed for work.   Suppose one took the system from state 1 to state 2 via path B and returned to state 1 by the reverse of path A. The value of

$$W = \oint dW = -\oint p_{ext}\, dV \qquad (5\text{-}16)$$

for this closed path is the negative of the area inside the cycle (single diagonal hatchmarks) and it is obviously not zero.   It could be made n times as big simply by repeating the cycle n times.   Its sign could be changed by taking the cycle in the opposite direction.   Thus $\oint dW$ can be made to assume *any* desired value, since it describes any path which begins and ends with the system in the same state.   Of course, if the path back to state 1 from state 2 is exactly the reverse of the outward path, then W for the cycle is zero. Different authors make different decisions about what they call simply "the work, $W = \int dW$."   Some choose, as we have, the work done on the system by its surroundings; some, however, choose the work done *by* the system.   This is one of several conventions an author of material on thermodynamics may use, and readers should always check his conventions when they start reading his work.   The two choices of W will almost always be negatives of each other, except for possible confusion when there is something analogous to friction.

To illustrate this confusion,* consider the expansion of a gas contained in a cylinder by a piston which moves uniformly and slowly outward, as shown in Fig. 5-7.   If there is no friction of the piston against the cylinder walls,

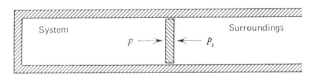

Fig. 5-7. Cross section of a cylinder of gas contained by a piston moving to the right.

the pressures are nearly equal across the piston and

$$p_{ext} = p_s = p \qquad (\text{no friction}), \qquad (5\text{-}17)$$

where p_s is the pressure to the right of the piston and p is the pressure in the

---

* Please note that the remainder of this section is for further clarification of this subject and may be omitted at the reader's discretion.

gas to its left.   If there is friction, however, the analysis depends on the definitions chosen.

If we treat the piston as part of the system, then the friction occurs inside the system.   The external pressure that enters the expression

$$dW = -p_{ext} \, dV \qquad (5\text{-}18)$$

is simply p_s, but this is less than the pressure p by an amount f/a,

$$p_{ext} = p_s = p - \frac{f}{a}, \qquad (5\text{-}19)$$

where f is the force of the frictional drag and a is the cross-sectional area of the piston.

If we treat the piston as part of the surroundings, the pressure at the left-hand side of the piston is p_{ext} and is equal to p.   However, it is greater than p_s by an amount f/a:

$$p_{ext} = p = p_s + \frac{f}{a}. \qquad (5\text{-}20)$$

With this definition, we note that a different value is found for dW by using Eq. 5-18.   This is because the process occurring in the immediate surroundings is complicated by the friction, the proper analysis of which is essential.

With both of these choices of definition the work done *by* the system is just the negative of the work done *on* the system.   However, occasionally the piston is viewed as neither part of the system nor as part of the surroundings, but as a wall separating the system to its left and the surroundings to its right. In this case, p_{ext} is given by Eq. 5-19, and the work done *on* the system is

$$\int p_{ext} \, dV = \int p_s \, dV = \int \left(p - \frac{f}{a} \right) dV. \qquad (5\text{-}21)$$

The work done *by* the system in this case is

$$-\int p \, dV \qquad (5\text{-}22)$$

and the two are not negatives.   This is because with this choice of definitions another part of the universe, namely the piston, does work on its surroundings equal to

$$\int \frac{f}{a} \, dV. \qquad (5\text{-}23)$$

As long as one carefully avoids this last kind of definition, the work done on the system will always be the negative of the work done by the system.

We have here an illustration of a common situation—we can all agree on just what is going on physically, but there are many ways of defining the words used to describe it.   Heated arguments often develop in such cases, since different people often use the same words defined differently.   To avoid these arguments, it is best always to seek agreement on the underlying reality and only later to consider how it is most conveniently described.

PROBLEMS

5-1. Calculate the work W for a compression from V_1 to V_2 performed at a constant external pressure of p_0.   What is W for the expansion which is just the reverse of the above?

5-2. A particular compression involving p-V work has $p_{ext}(V)$ given by

$$p_{ext} = 6.00 \text{ atm} - (0.1 \text{ atm/l})V.$$

Compute W for this compression from 40 to 10 l.   Sketch the p verus V plot before calculating.

5-3. A system undergoes a cyclic path which forms an ellipse on a p versus V plot.   The horizontal radius of the ellipse is 20 l, the vertical radius is 6 atm, and the path is counterclockwise.   What is the work W done on the system?

5-4. Invent a reasonable path between states 1 and 2 in Fig. 5-6 which has (a) an extremely large (nearly infinite) positive value for W; (b) an extremely large (nearly infinite) negative value for W; (c) $W = 0$.

5-5. How might you prove the result, Eq. 5-6, for p-V work for any infinitesimal volume change dV in an *arbitrarily shaped* system?

 6   THE FIRST LAW OF THERMODYNAMICS

A scientific law is a generalization from experience important enough to be called a "law."   In some branches of science very minor empirical correlations are called laws, but in thermodynamics the term is kept for the most important generalizations on which the whole subject is based.   There has always been the feeling that, starting with the "laws of thermodynamics," one should be able to derive all of thermodynamics rigorously, with the injection of no further generalizations from experience.   This accounts for the addition of the conventional zeroth and third laws to the first and second laws when their need was recognized in establishing temperatures and absolute entropies, respectively.

However, unless thermodynamic theory is treated like an abstract exercise, physical content enters the subject at all stages. In this book physical experience has already been discussed in some detail, and definitions and generalizations have been based on that experience. In the basic theory of thermodynamics, some ideas are emphasized by being called laws. When this theory is used with specific experimental information, the consequences may be examined to get a feeling for the conditions under which the various results would be expected and for the expected accuracy. During the process one is almost certain to be impressed by the marvelous variety of results which can be predicted and correlated by starting with only such apparently meager material.

The first law of thermodynamics is an extension to thermodynamic systems of the energy function of mechanics and of the mechanical law of conservation of energy in systems with only conservative forces. We first consider the energy function as it enters mechanics. Mechanical systems have no heat interactions, nothing complicated like friction, and no "thermal" variables such as temperature. Suppose such a system initially in state 1 changes to state 2, and in the process the surroundings do work W on the system. Whatever kind of work this is—mechanical, electrical, magnetic, etc., or some combination—the work may be converted as discussed in Sec. 5 (see Fig. 5-2 and Eq. 5-1), so that the sole effect in the surroundings is the raising or lowering of a standard weight in the earth's gravitational field. The remarkable fact of nature (or law of nature, or hypothesis about nature which has been confirmed) is that *a given change of state* (1 *to* 2) *for mechanical systems always leads to the same change in height for the weight, whatever the kind of work involved.* This is simply a restatement of the law of nature on which our entire discussion of work in Sec. 5 was based. It was the foundation of our assertion that consistent expressions may be found for the various kinds of work and for the relative values of the units in which they are expressed. If work-only experiments which take the mechanical system from state 1 to state 2 are done over and over, even though they take different paths and involve different kinds of work, the sole effect on the system is its change from state 1 to state 2 and the sole effect in the surroundings, when converted to a height change for the standard weight, proves to be independent of the path. The height change is a function of states 1 and 2 only. This is illustrated in Fig. 6-1. This invariance permits the definition of a function of the state of the mechanical system, which we call the **energy**, as follows: Start with the system in state 1 to which is assigned the arbitrary reference energy E_1. Take the system along any mechanical (that is,

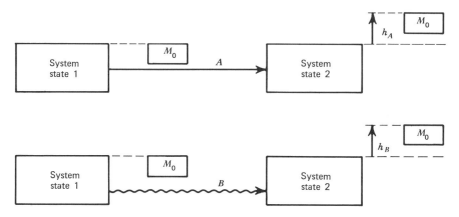

Fig. 6-1. Two different work-only processes (no heat), A and B, which take a mechanical system from state 1 to state 2 with sole effect in surroundings being the change in height of a weight. Whatever the paths, $h_A = h_B$.

work-only) path from state 1 to state 2 by using a converter which turns the effect of the work interaction in the surroundings into simply the change in height of the standard weight. Every state of the system reachable from state 1 has a unique value of the change of height h; therefore h is a property of the state of the system. Instead of using the height as the value of this property, we conventionally use the value of the work $-M_0 gh$ instead:

$$E_2 - E_1 = \int_1^2 dE \equiv -M_0 gh = -\int_1^2 M_0 g \, dh$$

$$= \int_{1 \text{ any path}}^2 dW = W \qquad \text{(mechanical systems).} \qquad (6\text{-}1)$$

The "energy of state 2 relative to state 1" for mechanical systems is simply the value of the work W done by the surroundings in taking the system from state 1 to state 2. Any mechanical path going from 1 to 2 may be chosen for finding the energy, because they all are found to yield the same work.

It is worthwhile considering why an extension of the energy function to thermodynamic systems might pose a problem, even if one rules out non-mechanical interactions like heat. If mechanical work is performed on a mechanical system, say, by pushing against a mass, the effect of the work is obvious in the mechanical response of the mass—it is accelerated to a new velocity. Similarly, if one does electrical work on an electric system, say, by charging a capacitor, the effect of the work is obvious in the electrical properties of the capacitor—the potential across it is changed. However, if one does mechanical work on a thermodynamic system, things might differ. Here, our pushing against a mass might be simply a rubbing against it with no resulting acceleration whatsoever; all the work is lost to friction. Similarly, electrical work can be done on a thermodynamic system in the manner of Fig. 5-5a, with no change at all in the electrical properties of the system.   Of course, the system will get warmer; and, in fact, the relationship between this warming and the work done is one of the things we want to study.

Mechanics restricts itself to systems for which the state is described by simple mechanical variables and among which the only interactions are work.   If we treat thermodynamic systems, we obviously must include non-mechanical or thermal variables to describe the system.   But we can restrict our considerations for the time being to adiabatically enclosed systems. For them, so far as communication between system and surroundings is concerned, the situation is the same as in mechanics.   We then simply look to experiments to determine whether an energy function of state defined by the amount of work exists for such adiabatically enclosed systems.   The answer is yes: *If an adiabatically enclosed thermodynamic system changes from state 1 to state 2 and the sole effect in the surroundings is the change in height of a weight, then the height change depends on only the initial and final states and not on the path.*   This is the **first law of thermodynamics**, and it may be written mathematically as

$$E_2 - E_1 = \int_1^2 dE \equiv -M_0 gh = -\int_1^2 M_0 g\, dh$$

$$= \int_{1 \text{ adiab path}}^2 dW = W_{\text{adiab}} \qquad \text{(all systems)}, \qquad (6\text{-}2)$$

$$\boxed{dE = dW_{\text{adiab}}} \qquad \text{(all systems)}. \qquad (6\text{-}3)$$

Thus the energy of even a thermodynamic system in any state may in principle be found relative to the energy in a reference state as long as *some* adiabatic (work-only, no heat) path exists which couples the two states.   The path need not proceed from the first to the second state; it might go from 2 to 1; one can get $E_2 - E_1$ simply from $-(E_1 - E_2)$.

One might ask if *any* two states of a system can be coupled by *some* adiabatic path.   Of course, until Sec. 12 we are restricting ourselves to cases in which the amount of material in the system stays constant.   Even then one cannot necessarily go *both* from state 1 to 2 and from 2 to 1 adiabatically. Consider the example of a gas with p_{ext} plotted against V, as in Fig. 6-2.   It might not be too hard to do adiabatic expansions and compressions $1 \to x$ and $y \to 2$ in either direction as indicated by the arrow heads.   However,

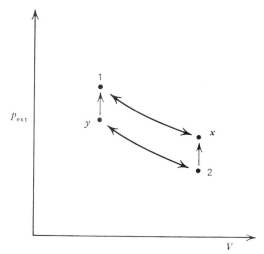

Fig. 6-2. Properties of a gas adiabatically enclosed.

the only adiabatic paths connecting y to 1 and 2 to x are upward as shown (this one-sidedness is discussed in Sec. 11.)   Such paths could be the performance of electrical work on the system by a scheme such as that of Fig. 5-5a (colloquially speaking, we would be using electrical work to "heat up the system").   However, this is enough to find the energy of any of the four points relative to any other, since energy is a function of state:

$$E_2 = E_1 + W_{1 \to x} - W_{2 \to x} = E_1 - W_{y \to 1} + W_{y \to 2}. \qquad (6\text{-}4)$$

Sometimes it is possible to measure heat just as precisely as work; these measured heat interactions can be as useful as work in finding ΔE for certain processes.   Therefore it is almost never necessary to worry whether two states can be coupled through adiabatic processes in order to know the difference in their energies.

Note that neither mechanics nor thermodynamics says that, given state 1 and W_{adiab}, there exists a unique state 2 in which the system will be found. That is obviously not true.   Given state 2 and W_{adiab}, one knows the energy of state 2, but there may well be an infinite number of states with the same energy E_2.

Since the energy of a system is a function of its state, the integral of dE around any closed path $\oint dE$ must be zero.   This means that, if an adiabatically enclosed system undergoes any process which returns it to its initial state and the sole effect in the surroundings is the change in height of a weight, then the height change must be zero.   If the height change were positive, we would have a device which could create potential energy out of nothing, a so-called *perpetual-motion machine of the first kind* (of the first kind because it violates the first law). Such a machine was eagerly sought for centuries; it would be nice to have one. Most people have given up looking; better uses of their time are not hard to find. If the height change were negative, we would have a sink for energy, again something very useful at a time when industries are hard pressed to find cheap ways to dissipate waste thermal energy without disrupting the ecological balance of neighboring regions. Again, all of man's scientific and technological experience tells him that such an energy sink cannot exist.

The desire of scientists to preserve energy conservation is responsible for the recognition of one new kind of energy after another, some of them most unexpected. For example, oil-soaked rags in air, adiabatically enclosed in a room, get hotter and hotter. How can one account for this apparent violation of the conservation of energy? Obviously the recognition of chemical energy released in the oxidation of the oil explains the phenomenon and preserves the concept of energy conservation. Another example is radium which gets hotter when adiabatically enclosed, even if surrounded by a vacuum. Since energy conservation makes sense out of *so many* phenomena, it can be abandoned only with the greatest reluctance. This explains the great attractiveness of Einstein's hypothesis that *mass* must be added (with the appropriate proportionality constant c^2) to the total energy of a system, a hypothesis which, of course, was precisely borne out by experiment.

Finally let us consider what is meant by "the energy of a composite" consisting of two subsystems, 1 and 2. The "energy" E_1 of subsystem 1 is defined relative to some reference state of subsystem 1 which is arbitrarily assigned the energy value of zero. The same is true of E_2 for subsystem 2. Now, if the subsystems which are not interacting with each other are viewed as a composite, we first must decide what reference state for the composite will have zero energy. The obvious choice is the state in which each subsystem has zero energy. With that choice, since work done on either subsystem changes the energy of the composite in the same way it changes the energy of the subsystem, *the energy of the composite is just the sum of the subsystem energies*:

$$E_{comp} = E_1 + E_2. \qquad (6\text{-}5)$$

Clearly, if the subsystems do interact with each other, it may be necessary to add an interaction energy to Eq. 6-5.

SUGGESTED READING

Feynan, Richard, *The Character of Physical Law*, The M.I.T. Press, Cambridge, Mass., 1965, chapter 3.

PROBLEMS

6-1. Do things happen because it is a law of nature that they happen, or is it a law of nature that they happen because things happen?

6-2. The first law is sometimes stated as "energy can neither be created nor destroyed." What more is needed to give meaning to this statement?

6-3. The first law is sometimes stated as "the energy of the universe (or world) is constant." In addition to the question raised in Prob. 6-2, what fault can be found with this statement?

6-4. A thermodynamics text says: "We can categorically deny the possibility of existence of phenomena (either at the scale of our planet or at a cosmical scale) in which the law of conservation and transformation of energy be invalidated." What can one conclude about such a statement?

6-5. Devise several devices which, if they worked, would violate the first law.

7   HEAT

So far we have not defined heat, but have just relied on its intuitive meaning. It is not easy to monitor just what is happening between two objects interacting through a diathermal wall. To illustrate this, we can consider three different kinds of diathermal walls, three ways in which heat interactions may occur:

The first is by *electromagnetic radiation* passing either from surroundings to system or vice versa, (actually, some radiation always passes in both directions). For example, the diathermal wall might be a cool region between a roaring fire and a sleeping dog. The dog's fur could be singed owing to electromagnetic radiation from the fire passing through the otherwise cool air to the dog. The space surrounding the earth is a diathermal wall through which radiation from the sun's hot surface is received and through which the earth, in turn, radiates and reflects radiation out to the rest of the (cold) universe. Not all radiation behaves like a heat interaction, sometimes it has the character of work. But those are special cases which involve either monochromatic radiation or collimated radiation, and they are best postponed. It is not easy to imagine a convenient way to *measure* quantitatively the amount of radiation passing through a diathermal wall.

A second way by which heat interactions may occur is by **convection**, that is, due to the bulk motion of material within the diathermal wall. For example, the wall might be the room full of air separating an iron " radiator " from the thermometer on the wall of a room. Heat from the radiator warms the air near it, which causes currents to be set up in the room, which eventually make the warm air circulate over the entire room, thus warming the thermometer. Again, it is hard to think up a straightforward way to measure quantitatively the " heat absorbed by the thermometer " from its surroundings.

A third way heat interactions may occur is by **conduction**, that is, a heat interaction through a diathermal wall, but caused neither by electromagnetic radiation nor by bulk motion of material within the wall. For example, an egg contained in a metal pan changes its properties dramatically when put over a fire as a result of heat conducted through the diathermal walls. Or, we might say, when we put our hand accidentally on a hot stove, that the diathermal wall (in this case, the imaginary boundary, or extremely thin surface region between hand and stove) conducts heat from the stove to the hand. (We are more likely to say something else!) In any event, the direct quantitative measurement of the " heat absorbed " in these interactions is almost inconceivable. What could one put in the bottom of the frying pan that would monitor the conduction process and yield a numerical measure of the heat absorbed by the egg?

Most real diathermal walls work in a combination of ways, and there are other kinds, such as so-called " heat engines," which are more complex. If the wall is a perfect vaccuum, radiation may be the only process, but material walls always involve both conduction and radiation, and fluid walls almost always involve convection as well.

How, then, might one define the heat absorbed by the system in the course of a particular process? Suppose the system, which is initially in state 1, undergoes a process involving a heat interaction with the surroundings and the performance of work W on it by them; the process concludes with the system in state 2. This is illustrated in Fig. 7-1. In principle, there is no difficulty in measuring the energy difference $E_2 - E_1 = \Delta E$ between states 2

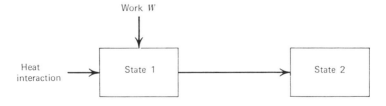

Fig. 7-1. System undergoing a process involving both work and heat.

and 1 by the methods of Sec. 6.   An adiabatic path is found which couples the two states, and the work performed on the system in the adiabatic process is ΔE.   This is sketched in Fig. 7-2.   Since energy is a function of state,

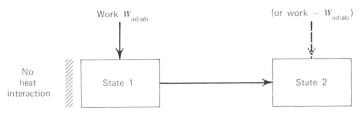

Fig. 7-2. Adiabatic process involving same change in state, performed to measure $W_{adiab} = E_2 - E_1 = \Delta E$.

ΔE is the same for the adiabatic path as for the path of Fig. 7-1, since both paths couple the same two states.   However, since work depends on the path, the W_{adiab} of Fig. 7-2 will in all likelihood differ from the value of W in the process of Fig. 7-1 where there is a heat interaction.   We *define* the **heat** Q absorbed by the system in a given process to be the increase in energy of the system due to the thermal contact with the surroundings.   As such, it must be the difference between ΔE and W, which is the increase in energy due to work for the particular process

$$\Delta E = Q + W. \tag{7-1}$$

For infinitesimal processes this becomes

$$dE = dQ + dW, \tag{7-2}$$

which yields Eq. 7-1 on integration:

$$\Delta E = E_2 - E_1 = \int_1^2 dE = \int_{1\ path}^2 dQ + \int_{1\ path}^2 dW$$
$$= Q + W. \tag{7-3}$$

Since W depends on the path and ΔE does not, Q is path dependent, dQ is not a perfect differential, and the same precautions must be taken for heat as for work.   Heat refers to a type of interaction between system and surroundings in which the internal energy of one is increased while that of the other is decreased by an equal amount (for proof see Prob. 7-1).   Heat is not "contained" or "stored" in a system; energy may be consistently spoken of in those terms, but neither heat nor work may be.   This requires a change in habitual thinking on the part of most people, because they are used to thinking about "the amount of heat in a body" when they mean the amount of thermal energy.

Equations 7-1 and 7-2 are concise statements of the first law of thermodynamics which hold for any process in which the amount of material in the system is fixed. They might make heat look like a fudge factor introduced to preserve energy conservation in thermodynamic systems. Actually, though, this is the only consistent definition of heat that follows from the first law. Sometimes heat is as easy to measure as work by noting the change in the system's surroundings. For example, consider a composite system, isolated from the rest of the world, consisting of two subsystems which we call the "system" and the "heat source." Suppose there is a heat interaction between them, as indicated in Fig. 7-3. If work is done by the surroundings just to restore the

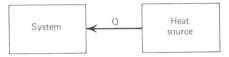

Fig. 7-3. Subsystems interacting through a diathermal wall.

heat source to its original state, the amount of work needed must be the same as the heat absorbed by the system. This is because ΔE for the heat source is zero if it is returned to its original state. This is a simple case of using work to restore a subsystem to its original state after it has experienced a heat interaction.

One frequently hears that "work is converted into heat" in slightly different situations where the statement is inaccurate. In Fig. 5-5b, where the resistance wire is part of the surroundings, electrical work done on the wire does "heat it up," with a resulting absorption of heat by the system. Work is indeed converted into heat. However, in Fig. 5-5a, electrical work is done directly on the system, increasing its thermal energy directly, with no heat involved at all. The process could, in fact, be an adiabatic one. The same is true of Joule's famous paddle-wheel experiment of 1840. First, the properties of a measured amount of water are used to define heat units, for example, calories or British thermal units. Second, a known amount of mechanical work is performed on the water to increase its thermal energy. Simply observing the energy change for the water as computed in heat units gives the conversion for energy changes via either heat or work. This is the so-called "mechanical equivalent of heat." Since then, however, the *thermochemical calorie* has simply been defined to be equal to 4.1840×10^7 ergs = 4.1840 joules. Much linguistic confusion stems from the colloquial usage of the word "heat" to mean thermal energy, that is, to say a hot object "has more heat" than a cold one or that "heat flows" from hot to cold objects. However, since these expressions confuse heat (which is descriptive of a process, that is, of an energy *transfer*), with energy (which is descriptive of a state), they should be avoided as much as possible.

In some complicated processes, even the distinction between work and heat may be impossible to make (see Bridgman, chapter 1). Then Q and W may have no individual meaning. However, ΔE will still be meaningful, as long as at least one path with well-defined Q and W can be *imagined* that couples the initial and final states. Even this may not be necessary (as Bridgman points out), so it is almost always possible to find the energy difference between two states. Where it is impossible, the very concept of the energy difference between the states would probably be meaningless.

We conclude by noting that by definition adiabatic processes have

$$Q = 0 \quad \text{(adiabatic process).} \tag{7-4}$$

Processes occurring in isolated systems have

$$Q = 0, \quad W = 0, \quad \Delta E = 0 \quad \text{(isolated system).} \tag{7-5}$$

SUGGESTED READINGS

Bridgman, P. W., *The Nature of Thermodynamics*, Harper & Brothers, New York, 1961, chapter 1.
Callen, H. B., *Thermodynamics*, John Wiley & Sons, New York, 1961, pp. 19–20 (for an interesting analogy to the concept of work and heat).
Van Ness, H. C., *Understanding Thermodynamics*, McGraw-Hill Book Company, New York, 1969, chapter 1 (in paperback).

PROBLEMS

7-1. Prove that, in a heat interaction between objects A and B through a diathermal wall, the energy of one is increased while that of the other is decreased by an equal amount.

7-2. How could you measure the heat absorbed by an object during its brief insertion into a very hot furnace?

7-3. Suppose a gas is contained in a metal cylinder by a piston, and the whole thing is suspended in a large lake. The surroundings do work on the gas by very slowly compressing it, at the expense of the lowering of a weight. No other change is detectable in the surroundings. Is the energy of the gas at each stage of the compression simply the work done? Why?

7-4. In what way would there be heat and in what way work interactions in the following processes (*a*) A stopcock is opened, permitting the gases in two connected flasks to mix; (*b*) the air in a bicycle tire pump is squeezed into a tire; (*c*) a mixture of H_2 and O_2 is exploded by a tiny spark in a flexible container at 1 atm of pressure; (*d*) the same mixture is exploded in a strong steel vessel; (*e*) a liquid swirling around in a jar comes to rest through viscous forces; (*f*) a pressure cooker containing water, meat, and air is put on a stove with escape vent open; (*g*) same as (*f*) with escape vent sealed.

7-5. A heat engine is a device which absorbs heat $|Q_1|$, discharges heat $|Q_2|$, does work $|W|$ on the surroundings, and returns to its original state. Express the *efficiency* of the engine, defined to be the fraction of the heat absorbed that is given off as work $|W|/|Q_1|$, in terms of $|Q_1|$ and $|Q_2|$. *Note*: The use of $|\ \ |$ around a quantity means its absolute value, regardless of sign.

8 THE SECOND LAW OF THERMODYNAMICS

Every day we experience the fact that systems left to themselves run down and the macroscopic properties that describe them become independent of time: Friction damps mechanical motion in even the most frictionless machines; heat flows from hot regions to cold until both regions have similar properties; a mouse isolated inside a bell jar eventually dies; matter diffuses from regions of high concentration to regions of low until the concentrations become uniform; a watch left to itself runs down and quits ticking; viscous drag arrests the motion of particles in fluids and damps out currents in the fluids themselves; a ball dropped on a pavement ultimately stops bouncing; and a mixture of natural gas and air burns to give water and carbon dioxide and then the reaction stops.   Such processes lie at the heart of thermodynamics in the generalization called the second law.

The **second law of thermodynamics** states that *the macroscopic properties of any isolated system eventually assume constant values.** Since an isolated system is said to be in **equilibrium** when for all practical purposes its thermodynamic properties do not change with time, the second law simply states that any isolated system evolves until it reaches an equilibrium state.

Nothing in the second law says anything about *how fast* the approach to equilibrium will be.   Some chemical reactions reach equilibrium in 10^{-6} sec, an ice cube put into a glass of water melts in a few minutes, a well-made pendulum might swing for weeks before its oscillations are damped, an iron nail takes years to crumble into rust, and some reactions like the conversion of diamond to graphite go so slowly that even in geological times their

---

* This is not the conventional statement, but we like it better.   It is slightly broader and considerably more intuitive.   From it we shall deduce the conventional zeroth law and the conventional statements of the second law due to Kelvin, to Clausius, and to Caratheodory.

progress is negligible.   The time it takes for a process to reach equilibrium is commonly called the **relaxation time** for the process.*   Classical equilibrium thermodynamics directs its attention to ultimate equilibrium states, that is, it looks at times which are long compared to the relaxation time.   If experimental observation times are about the same as the relaxation time, then conventional nonequilibrium thermodynamics or chemical kinetics might be useful in studying the approach to equilibrium.   However, if observation times are much shorter than the relaxation time, then the best thermodynamics can do is to ignore that relaxation process and treat it as being, for all practical purposes, frozen.   The function of a *catalyst* is to change the relaxation time, sometimes by factors of 10^{100}.   Negative catalysts increase it, positive ones decrease it.

The second law holds only for macroscopic variables, for which significant fluctuations, as discussed in Sec. 2, may be neglected as too incredibly unlikely.   The second law is violated by fluctuations away from the stationary values of the equilibrium properties of a system.   The problem of fluctuations and the second law is interesting, and there have been many fights over it; some still go on.   Often they are fights over definitions.   It was proved by Henri Poincaré in 1890 on the basis of molecular theory that no matter how wild and unlikely a fluctuation is, if one only waits long enough, *it will occur*!   Any state that it is possible for the material of a system to get into will eventually be reached by that system.   Suppose the two bottles in Fig. 8-1 contain

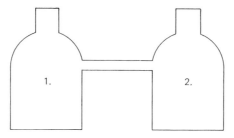

Fig. 8-1. Two bottles containing gas.

a dilute gas.   Poincaré's proof implies, for example, that if one waits long enough he will find all the gas in bottle 1 (or bottle 2).   This is really a proof that *the second law as we have stated it is false!*   But what are the practical consequences of its falseness?   How false is false?   If the probability that any one gas molecule be in bottle 1 is $\frac{1}{2}$, then the probability that all N of them be there is $(\frac{1}{2})^N$.   If 10 molecules were in the two bottles, this is a

---

* Or, more precisely, the relaxation time is the time a process takes to go $1/e$ of the way from the starting state to the final equilibrium state, where $e = 2.72$ is the base of the natural logarithms.

probability of 10^{-3}, so one would probably see the fluctuation if one looked 1000 times.   All 10 molecules just might be found in one bottle or the other. If 100 molecules were in the system, the probability that they are all in the first bottle is 10^{-30}; thus if one looked at the bottles once each second, one might expect to find all of the 100 molecules simultaneously in one bottle if one looked for 10^{12} times as long as the estimated age of the universe!   With this kind of probability for a wild fluctuation involving only 100 molecules, just think of how long one would have to look in order to expect to see such a wild fluctuation if N were 10^{20} or so!   This is why macroscopic variables as discussed in Sec. 2 are never seen to violate the second law; big fluctuations *just do not occur*, even though they certainly *would* occur if one could just wait long enough.

We have defined the equilibrium condition only for isolated systems, but systems contained within diathermal walls can reach equilibrium too.   A system in thermal contact with its surroundings is in equilibrium when (*a*) for all practical purposes its thermodynamic properties do not change with time and (*b*) it and the entire surroundings with which it is in contact are capable of being approximated by a single isolated equilibrium supersystem.   The reason for this latter proviso is to prevent using the word *equilibrium* to describe what should be called a nonequilibrium *steady state* maintained by the surroundings.   For example, if chemical reactants are fed into a reaction vessel at constant rates, and products are removed at constant rates, the state of the system may become time independent, but this is certainly not equilibrium.   Similarly, a system making thermal contact on one side with an ice bath and on the other side with a steam bath may reach a time-independent condition.   However, this steady state is not equilibrium, since the system, ice bath, and steam jacket taken together could not be approximated by a single equilibrium supersystem.*

SUGGESTED READING

Feynman, Richard, *The Character of Physical Law*, The M.I.T. Press, Cambridge, Mass., 1965, chapter 5 (paperback).

---

* It would seem attractive to postulate a nonequilibrium law of the steady state, analogous to the second law, namely, that the macroscopic properties of any system exposed to time-independent surroundings will eventually assume constant values.   However, such a postulate is demonstrably untrue.   A simple example is the fact that fluid forced steadily through a channel develops turbulent flow in which the local velocity assumes a variety of interesting values rather than a steady value.   The emergence of life on the crust of the earth under the steady driving force of the absorption of high-temperature radiation from the sun and the emission of medium-temperature radiation to the rest of the universe is another example of a most interesting violation of the inevitability of the steady state!

PROBLEMS

8-1. In their book, *Statistical Mechanics*, John Wiley & Sons, New York, 1940, J. E. Mayer and M. G. Mayer introduced the second law of thermodynamics by quoting from Gilbert and Sullivan's *H.M.S. Pinafore*: "What, never?" "No, never!" "What, never?" "Well, hardly ever!" Why was this relevent to the second law?

8-2. Propose several devices which, if they worked, would violate the second law but not the first.

8-3. A thermodynamics text says: "The 'theory' of the thermal death of the universe leads directly to religious superstition—to the existence of God. In fact, since according to Clausius the universe moves continuously towards thermodynamic equilibrium, while it is not in an equilibrium state at the present moment, then it follows that either the universe has not always existed and has been somehow created, or some sort of external force has at some time removed it from its equilibrium state and we are now living the epoch of the returning of the universe to a state of equilibrium. This means that God must exist." Discuss.

9 THERMAL EQUILIBRIUM AND TEMPERATURE

ZEROTH LAW OF THERMODYNAMICS

This book has been organized so that its first eight sections are completely general; they apply to all systems, whether in equilibrium or not. The rest of the book pertains only to equilibrium states and the processes coupling them. Only after understanding equilibrium does it pay to worry about the rates of nonequilibrium processes.

The first law postulated (or generalized from experience) the existence of a function of state—the energy—and prescribed how to measure it. From the law of equilibrium that we have called the second law we may deduce the more conventional statements of the second law and also the so-called "zeroth law." The conventional zeroth law postulates (or generalizes from experience) the existence of another function of state—the temperature of an equilibrium system—and prescribes how to measure it. The conventional second law similarly gives rise to the closely related function of an equilibrium state—the entropy—tells how it may be determined, and tells how knowledge of a variety of thermodynamic functions may be computed if the entropy is known. In this section temperature is discussed; entropy is developed in Sec. 11.

Suppose one connects any two objects, whatever their properties, by a diathermal wall. They do no work on each other, but heat flows from one to the other until, finally, thermal equilibrium is reached. The properties of the objects become time independent in accordance with the second law. We say these two objects then have the same "temperature." Two subsystems are said to be **at the same temperature** whenever they are in thermal equilibrium, that is, when they are in equilibrium through a diathermal wall.

We now consider whether it is possible to establish various "empirical temperature scales" through which numerical values may be assigned to the temperature as a function of state in a consistent way. For such assignments to be consistent, it is necessary that (1) all objects with the same value of the temperature should already be in thermal equilibrium with each other whenever they are placed in thermal contact, and (2) it should be possible to assign temperatures consistently so that objects "hotter" than others have higher temperatures than the others. If either of these requirements could not be met for the objects of our experience, then the notion of temperature as a function of state would be useless.

In order to deduce these two important requirements from our statement of the second law, we need to admit the existence in nature of so-called "heat engines," which have the following properties: A **heat engine** is a device which has heat-only interactions with two or more objects and which, in consequence, does work spontaneously. At the end of a period of operation, the heat engine itself is in the same state in which it started out. The work may be done either on the surroundings or back on one or both of the objects. Objects, themselves internally nearly in thermal equilibrium, used like this to supply or absorb thermal energy in the form of heat are called *thermal reservoirs*, *heat reservoirs*, or *heat baths*. The action of a heat engine is sketched in Fig. 9.1.   Heat engines may be used as a special class of diathermal walls by simply letting two reservoirs at different temperatures approach thermal equilibrium through the heat engine while the engine does its work back on the subsystems (so the energy of the composite remains unchanged). This is sketched in Fig. 9-2.   In the course of this work, the heat engine may move around, since work in any form can always be converted into mechanical work.   Heat engines, from this viewpoint, are the class of diathermal walls which may move as the subsystems approach thermal equilibrium through them.

Obviously, such devices exist.   The steam engine is a famous example in which mechanical work is done through heat interactions of the steam with

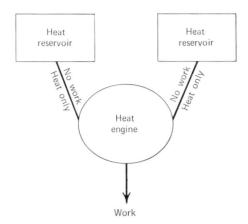

Fig. 9-1. A heat engine in operation.

the hot boiler and the cool surrounding air. Another good example is the thermocouple, a device which directly connects hot and cold metal-to-metal junctions and generates an electric voltage thereby, thus doing electrical work as desired. Certainly the most interesting heat engine is the surface of the earth, the biosphere, which has a heat interaction via radiation with the hot surface of the sun and has another with the cold remainder of the universe (at an average temperature of a few degrees Kelvin), in the "moving around" of which all life has risen.

Different heat engines show differing sensitivities to states of thermal disequilibrium. This is analogous to the fact that different mechanical devices, when acted on by driving forces, require different forces before they can overcome their internal friction and begin to move. A steam engine obviously requires a large thermal disequilibrium before it will begin to move, but some

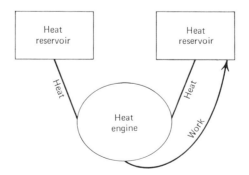

Fig. 9-2. Heat engine used as a diathermal wall.

extremely sensitive thermocouples will start to do electrical work whenever the heat reservoirs to which they are coupled are changed only infinitesimally in one way or the other from a state of equilibrium. It is convenient to use such a very sensitive heat engine as a test device for determining whether two subsystems are in thermal equilibrium.

Granted such heat engines exist and that isolated systems inevitably approach equilibrium, we can immediately deduce that *any two systems found to be in thermal equilibrium through the most sensitive of heat engines must prove to be in thermal equilibrium through any other diathermal wall or heat engine.** The converse must also hold: *any two systems in thermal equilibrium through any diathermal wall must prove to be in thermal equilibrium through any heat engine, even the most sensitive, and consequently in thermal equilibrium through all other diathermal walls.* This is clarified by Fig. 9-3,

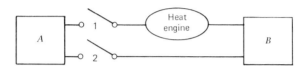

Fig. 9-3. Systems A and B able to make thermal contact via switch 1 through a sensitive heat engine or via switch 2 through a diathermal wall.

in which two systems are able to be placed in thermal contact either through a sensitive heat engine or through some other diathermal wall. The two switches can be thrown in principle with a negligible amount of work. If switch 1 is closed, A and B will reach equilibrium through the heat engine, which can do its work back onto one of them in the manner of Fig. 9-2. If now switch 1 is opened and switch 2 closed, A and B will be in thermal equilibrium, or else heat will flow through the diathermal wall connecting them and they will change their states. If that happened, switch 2 could be opened and switch 1 closed and the heat engine would run again. This process could be continued, switching from one path to the other with the engine running intermittently, the whole thing driven spontaneously and going on forever. In that case, the entire contents of Fig. 9-3 would represent an isolated system which refused to run down and reach equilibrium. This is impossible, so we have deduced the statements that began this paragraph: *Every conceivable kind of diathermal wall, including all kinds of heat engines, must agree on what are states of thermal equilibrium.*

* Or course, the wall itself must have negligible *direct* effect on the subsystems due to its thermal energy or lack thereof. A block of ice used as diathermal wall obviously affects the final equilibrium differently from a container of boiling water, since their heat capacities are not negligible. We are not talking about this type of direct effect of the wall itself.

As noted above, if meaningful temperatures are to exist, the first require-ment is that *things in thermal equilibrium with the same thing will be found to be in thermal equilibrium with each other*. This is what is conventionally called the **zeroth law of thermodynamics**.   It is a statement about the experi-ment illustrated in Fig. 9-4.   Suppose system *A* is in thermal equilibrium with

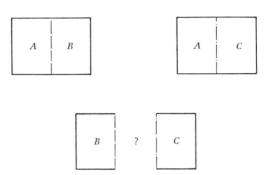

Fig. 9-4. System *A* in thermal equilibrium with *B*.   Also *A* with *C*.   What about *B* with *C*?

system *B*.   Suppose also that *A* in the same state is found separately to be in thermal equilibrium with system *C*.   We would say from the first experiment that *A* and *B* "have the same temperature" and from the second experiment that *A* and *C* do too.   Clearly, if temperature is meaningful, it must follow that, when *A* is connected by a diathermal wall to *C*, the properties of *A* and *C* must not change, that is, they must already be in thermal equilibrium since they have the same temperature.

The conventional statement of the zeroth law is a corollary of what we deduced above, since Fig. 9-4 is completely equivalent to Fig. 9-5.   Here it

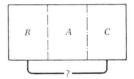

Fig. 9-5. Equivalent to Figure 9-4.

is clear that system *A* in Fig. 9-4 may, if we wish, be viewed as a *diathermal wall* coupling systems *B* and *C*, and we have already shown that *B* and *C* must prove to be in thermal equilibrium by whatever device is inserted at the question mark in Fig. 9-5.   Therefore we are able to say consistently that all systems which would be found to be in thermal equilibrium with each other if they were placed in thermal contact have the same **temperature**.

CLAUSIUS' STATEMENT OF THE SECOND LAW

We now want to go further and assign numerical values to the temperature as a function of state of the system.   How we assign these numbers is completely arbitrary; any choice we care to make defines an empirical temperature scale in which the *empirical temperature* θ is given as a function of the variables describing the state of the system.   There are two steps in constructing an empirical temperature scale.   The first is to choose quite arbitrarily some physical object called the *thermometer* and to choose some property of that object to measure the temperature.   Examples are many: the length of a mercury or alcohol column inside a sealed glass tube, the resistance of a piece of platinum or carbon wire, the pressure of some gas sealed into a certain volume, a plot of intensity versus wavelength for the radiation emitted by a glowing object, the magnetic moment of a paramagnetic substance placed in a given magnetic field, the voltage established by a thermocouple one junction of which is kept in an icebath, etc.   The thermometer is placed in thermal contact with the object the temperature of which is to be measured. The numerical value of the chosen property (or any simple single-valued function of that value) of the thermometer is the *empirical temperature* of the object.   Empirical in this case means arbitrary, dependent on the use of this particular thermometer and this particular scheme for going from its properties to a numerical temperature.   If the thermometer is mercury in glass, the empirical temperature could simply be the column length in centimeters.   If the thermometer is a piece of resistance wire, the empirical temperature could be its resistance in ohms.   Objects at the same temperature will always leave the thermometer in the same state.   Objects of different temperature leave the thermometer with different values of the chosen property, the difference being a measure of their thermal disequilibrium.

As noted at the beginning of this section, the second feature that any temperature scale should possess is that hotter objects should always have higher temperatures (or lower, but one or the other) than cooler ones.   By "hotter" is meant no subjective determination made by touching the objects, but the following experimental definition: If two bodies with different temperatures are placed in thermal contact, there will be a heat interaction until their temperatures become equal.   The heat Q will be positive for the body whose thermal energy increased, negative for the other.   The body with negative Q is said to have been the *hotter* of the two at the beginning of the experiment.   The possibility of choosing temperature scales so that $\theta_A > \theta_B$ always implies that A is hotter than B follows directly from the second law.   If this were not true, then three bodies could exist at equilibrium for which heat would flow from A to B, from B to C and from C to A. This is sketched in Fig. 9-6a. If this were possible, heat could be allowed to flow from A to B and from B to C, while the flow from C to A would be replaced

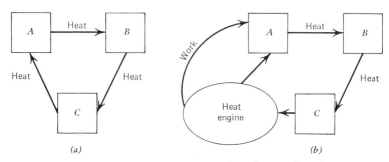

Fig. 9-6. (a) An (impossible) situation in which $\theta_A > \theta_B$ and $\theta_B > \theta_C$ does not imply $\theta_A > \theta_C$. (b) The harnessing of (a) to do mechanical work continuously in an isolated composite system, thus violating the second law.

by a heat engine acting between C and A whose temperatures would be different. Again, this composite could move around forever, a violation of the second law. Thus a consistent correspondence between temperature and hotness must be possible. This conclusion from our statement of the second law is so basic that it is, in fact, one of the classical statements of the second law, that due to **Clausius**: *It is impossible to have a natural process which produces no other effect than the absorption of heat from a colder body and the discharge of that heat into a warmer one.* This is clearly just a statement that all diathermal walls, however cleverly they may be constructed, must lead to the same thermal equilibrium; when hot and cold bodies are placed in thermal contact, the hot ones get cooler while the cold ones get warmer. If this were not so, then by using a heat engine, as we have seen, a perpetually moving device—above the level of Brownian motion and other fluctuations—could be made in an isolated system. Such a *perpetual-motion machine of the second kind* (since it violates the second law) disagrees with our experience. Of course, devices such as refrigerators can be made which transfer energy from colder to hotter bodies, but their operation is driven by work from outside. In order to accomplish this work, permanent effects such as the lowering of a weight in a gravitational field are made in the surroundings.

We now have deduced both requirements which must be features of the behavior of bulk matter if temperature scales can be established in consistent ways. The first requirement is the conventional zeroth law of thermodynamics and the other is conventionally called Clausius' statement of the second law. The fact that both are needed to establish empirical temperatures and not just the first is often overlooked by writers on thermodynamics. Both features were deduced from our general statement of the second law as a law of equilibrium, with the addition only of the properties of heat engines.

THE KELVIN TEMPERATURE SCALE

In fact, of course, empirical temperatures are not often recorded simply in inches of an alcohol column or in dynes per square centimeter pressure for a gas sealed in a flask, because it is so hard to reproduce exactly the same kind of thermometer from one place or time to another. Instead conventional empirical temperature scales are chosen with selected reference temperatures which can be reproduced anywhere or any time. Then local thermometers can be calibrated by comparison with the references and reproducibility becomes possible. A variety of conventional empirical temperature scales have been devised for use in various applications—Kelvin, Rankine, centigrade (or Celsius), Fahrenheit, and others. In basic science the Kelvin scale is by far the most important, so we discuss it here and view the others as derived from it by simple reassignments of numerical values.

The thermometer for the Kelvin scale is a measured number of moles of any real gas sealed into a chamber for which both pressure and volume can be measured. If the temperature of the gas is kept fixed by its being kept in thermal equilibrium with a large heat bath, repeated measurements of p and V at ever larger V (and thus ever smaller molar density, $\rho = n/V$) can be made. A plot can be drawn of pV/n (that is, of p/ρ) versus $\rho = n/V$, as in Fig. 9-7 for small ρ. Obviously it is impossible to perform pressure measure-

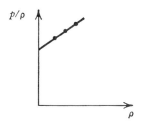

Fig. 9-7. Typical plot of p/ρ versus ρ for a real gas at small ρ and fixed θ.

ments on gas of zero density, but it is possible to do them on sufficiently dilute gases to warrant extrapolating the data to $\rho = 0$. This is shown in Fig. 9-7. The interesting feature is that the plot of p/ρ actually extrapolates to an intercept on the axis at $\rho = 0$, rather than oscillating or becoming tangent to the axis, as shown in Figs. 9-8. Therefore, the intercept does not depend on ρ, although, of course, it does depend on the value of the constant

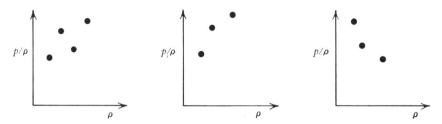

Fig. 9-8. Behavior of p/ρ versus ρ that is *not* observed for real gases at small ρ and fixed θ.

temperature θ.   Since the intercept is the limiting value of p/ρ as $\rho \to 0$, we we can write this experimental result thus:

$$\lim_{\rho \to 0} \frac{p}{\rho} = f(\theta \text{ only}) \qquad \text{(real gas)}, \tag{9-1}$$

where f is some function of the empirical temperature only.

One further interesting conclusion arises from these experiments on real gases, namely, that for a given temperature the value of the p/ρ versus ρ intercept is *independent of what gas is being studied*.   This is illustrated in Fig. 9-9, where data for several different gases are plotted on the same p/ρ

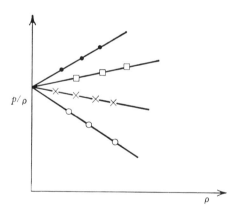

Fig. 9-9. Plots of p/ρ versus ρ for four different real gases at small ρ and the same fixed θ.

versus ρ graph, and each gas extrapolates to the same intercept for the particular temperature.   We thus conclude on the basis of experiment that, *in the limit of vanishing density, the pressure divided by the molar density of all real gases becomes a universal function of the temperature only*:

$$\lim_{\rho \to 0} \frac{p}{\rho} \equiv \lim_{\rho \to 0} \frac{pV}{n} = f(\theta \text{ only}) \qquad \text{(all real gases)}. \tag{9-2}$$

By **universal function** $f(\theta)$ is meant the same function for all gases of the particular empirical temperature chosen. The limit of p/ρ has the same value for CO_2, CH_4, Ar, Cl_2, Kr, He, C_2H_6, O_2, N_2, ..., provided they are all at the same temperature.

The only remaining task in using experimental determinations of the p/ρ intercept as a standard thermometer is to associate numbers with various values of the limit of p/ρ. It might have been convenient if the left-hand side of Eq. 9-2 had been defined as $1/\theta$, which would have given temperature the units of reciprocal molar energy and would have meant that heat flowed from regions of low to regions of high temperature. Another convenient choice would have been to define it to be θ directly, giving temperature the units of energy per mole. However, temperature scales were in use for centuries before it was possible to make as detached an overview as that presented here. The centigrade scale with $100°$ between the freezing and boiling point of water was the choice of most scientists. This fixed the unit of temperature as the degree and also fixed its size. Since one objective in defining the temperature scale is to minimize the need to revise tables of existing data, the absolute or Kelvin scale has been defined to preserve as nearly as possible the former " absolute temperatures." This meant defining Eq. 9-2 to be *proportional* to θ,

$$\lim_{\rho \to 0} \frac{p}{\rho} \equiv R\theta \qquad \text{(any real gas)}, \qquad (9\text{-}3)$$

with a proportionality constant R which is not itself defined, but which is fixed by a further experiment. Once Eq. 9-3 is written and the decision is taken to establish a new unit for temperature, the *degree*, then fixing the value of R fixes the size of the degree. This is done by defining one standard reference temperature, that of pure water at its triple point, to be $273.1600°$. It is not too hard to make up a reference system of pure water in which solid ice, liquid water, and water vapor are coexisting in equilibrium. There is only one unique state of temperature and pressure at which these three phases coexist, and it is called the *triple point* for water. Then, if any real gas is kept in thermal equilibrium with this reference system, the left-hand side of Eq. 9-3 can be measured by the extrapolation procedure. Since θ is then by definition $273.1600°K$, the value of R is determined. Experiments show that R, known as the **universal gas constant**, has the value of 0.08205967 l-atm/deg-mole $= 1.98726$ cal/deg-mole $= 8.31469 \times 10^7$ ergs/deg-mole.

With the value of R finally determined, Eq. 9-3 provides a means to fix numbers to any temperature by performing the extrapolation procedure for any real gas at the temperature of interest. The scale so defined is the **Kelvin temperature scale**:

$$\theta_{\text{Kelvin}} = \frac{1}{R} \lim_{\rho \to 0} \frac{p}{\rho} \qquad \text{(any real gas)}, \qquad (9\text{-}4)$$

with R equal to 0.08205967 l-atm/deg-mole, or an equivalent value in other units.   In terms of Kelvin degrees, the standard melting point of water (the zero of the centigrade scale) is 273.15°K.   Experiments with real gases can now be used to calibrate precise thermometers, such as the mercury-in-glass type, which are more convenient to use than gas thermometers.   Once this is done, these secondary thermometers may be used to measure temperatures, the propriety of the usage being guaranteed by the zeroth law of thermodynamics.

THE CELSIUS TEMPERATURE SCALE

Because of the great difficulty in performing experiments with dilute gases, the Ninth General Conference on Weights and Measures (1948) agreed upon an empirical temperature scale more readily amenable to experimental procedures.   It is the *international temperature scale of* 1948, designated degrees Celsius (°C) and given the symbol t (at least, where there is no chance of confusion with time).   The Celsius scale in the range -182.970 to 630.5°C is based on the resistance of a piece of platinum of specified design.   In the range 630.5 to 1063.0°C, it is based on the voltage developed in a standard platinum–platinum-rhodium–alloy thermocouple with one junction at the ice point (0°C) and the other at t.   For most practical calculations, the conversion between the Kelvin and Celsius scales is

$$\boxed{\theta_{\rm K}\,(^\circ K) = t\,(^\circ C) + 273.15^\circ} \tag{9-5}$$

although the constant 273.15° actually varies by several tenths of a degree over the range of some 800°.   For further details, see H. F. Stimson, *J. Res. Natl. Bur. Std.*, **42**, 209 (1949).

PROBLEMS

9-1. A. N. Sharpe has observed (*Chemistry and Industry*, Feb. 11, 1967, Letters to the Editor) that in the temperature range of 8 to 29°C the number of slices of bread which can be spread with $\frac{1}{2}$ lb of butter is equal to the temperature in °C If 1 lb spreads 40 slices, what is the temperature in °K and in °F?

10 REVERSIBLE PROCESSES

The second law, as we have stated it, formally stresses the irreversibility of natural processes. It asserts that no process is truly reversible, in that systems are always evolving toward equilibrium and away from states of disequilibrium, so the reverse of the natural process would be impossible. Nevertheless, it is most useful to define reversible processes, to study many of their features, and then to analyze many real processes in terms of reversible ones.

We define a **reversible process** to be *one each infinitesimal step of which may be exactly reversed by an infinitesimal change in the external conditions prevailing at the time of that step.* It is easy to give examples to show why certain processes are irreversible. Consider a volume of gas held by a piston inside a cylinder, and let the external pressure be plotted against volume, as shown in Fig. 10-1. Initially, let p_{ext} and V be given by point 1, and suppose that the pressure in the system is also p_1. Suppose now it is desired to increase the volume from V^0 to V'. The natural way to do this is to decrease p_{ext} until the piston begins to move out. We note the pressure at which this happens by p_2. The reason p_2 is less than p_1 is that the frictional force on the

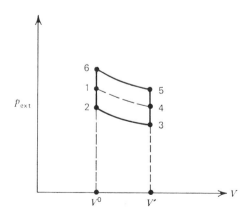

Fig. 10-1. External pressure plotted against volume of a gas.

piston in the cylinder must be overcome by the product of $p_1 - p_2$ times the cross-sectional area of the piston. When this friction is finally overcome, the piston moves and the volume can be increased to V'. When V' is reached, p_{ext} may be increased from p_3 to p_4, the new equilibrium pressure within the system, and the motion of the piston stops.

Now if one desires to return this system to its initial condition, one must increase the pressure to some higher value, p_5, to overcome friction, and then at point 6 return the pressure to p_1. The net work done by the surroundings on the system during the overall process is $-\int p_{ext}\, dV$, which is the area inside the figure 123456. This work has been converted into thermal energy. If the process had fit the definition of reversible, then $p_1 - p_2$ and $p_5 - p_4$ would have been truly infinitesimal, that is, capable of being made smaller than any arbitrarily chosen small values. The net work done by the surroundings on the system and turned into thermal energy during the overall process of expansion and subsequent contraction would then have been infinitesimal too. But we know this is impossible because any piston has a finite frictional drag at the cylinder wall.

Other effects also contribute to the irreversibility. Once the friction is overcome at points 2 and 5, the piston begins to move at a finite rate which sets up waves in the gas; that is, mechanical energy is given to the gas which ultimately is dissipated into thermal energy through viscous and turbulent processes. Reversal of this process is impossible, since under no circumstances will thermal energy spontaneously take the form of waves. Also, if the piston is moving at a finite rate at points 3 and 6, it must be given an abrupt impulse to halt it, and the spontaneous reversal of these impulses is inconceivable.

Thus one concludes that two features of the real world combine to prevent reversibility in the p-V process analyzed: these are *friction* and the *finite rate* of the process. Although these features are always present in mechanical processes, they can be minimized. Pistons can be made with less and less friction by careful machining and lubrication. The process may be performed more and more slowly, so that the duration of the process is much greater than relaxation times for any viscous or turbulent processes inside the gas, and so that the impulses needed to halt the piston at points 3 and 6 are negligible. If this procedure could be imagined being refined more and more, in the limit of perfection, a reversible process would be obtained.

We do not intend to sound overly pessimistic. For all practical purposes, a great many processes can be treated as reversible. Many which cannot are best handled as reversible processes, with corrections for friction and finite rates. The idea of the reversible process as the limit of natural, irreversible processes is also very useful in many theoretical discussions. Thus the great attention devoted to reversibility in thermodynamics.

Reversibility in heat interactions between system and surroundings also requires analysis. The temperature scale has been chosen so that heat "flows" from regions of higher to lower temperature. In order for an infinitesimal change in the surroundings exactly to reverse the direction of a small heat interaction, the temperatures of system and surroundings must differ by only an infinitesimal amount. This is the equivalent to the no-friction requirement for p-V work. It is also necessary for the process to go slowly enough so that, as the thermal energy enters or leaves the system, the temperature of the entire system remains uniform, just like it would be in equilibrium. The internal relaxation process here is governed by the thermal conductivity of the system. The larger the conductivity, the easier it is to maintain the equilibrium condition, and the faster may be the heat flow and still approach reversibility.

Other examples which prevent reversibility are overvoltages in passing electric current through systems and hysteresis effects in changing the magnetic fields on systems. In some systems which have very short internal relaxation times, processes have to be very fast indeed before the rate leads to appreciable irreversibility. For example, the direction of the magnetic field applied to a paramagnetic solid may be reversed every 10^{-9} sec without measurable introduction of irreversibility into the magnetization process. However, some processes, such as changing the shape of an almost infinitely viscous liquid (like glass), just cannot be performed slowly enough to begin to approach reversibility.

To summarize, there are **two requirements for a process to approach reversibility**: (1) *The change must be performed slowly* with respect to all internal relaxation times for the system, such as thermal conduction or viscous or turbulent relaxation. (2) *There must be no friction* (or overvoltage, or hysteresis, or whatever plays this role) *and no finite temperature differences*. Each process must be analyzed separately to see how closely it fulfills these criteria.

Reversibility can also be prevented simply by putting some kind of sensor on the system which prevents its going both ways. For example, a ratchet could be used to keep the piston from returning once it has moved out. Such a process might still have all the attributes of a reversible one in one direction, but simply is prevented physically from taking the reverse step.

Once a process is admitted to be reversible for all practical purposes, there are two immediate conclusions that can be drawn about it: (1) Since the change is performed slowly with respect to internal relaxation times, the process could be stopped at any stage and the system would be indistinguishable from one which had been at equilibrium for years. Thus the state variables describing a system undergoing a reversible process are related in exactly the same way they would be were the system in equilibrium. (2) The external forces are uniquely related to the internal forces, and indeed are usually equal. For example, if the p-V process of Fig. 10-1 were reversible, then

$$p_{ext} = p_{int} \pm \text{infinitesimals} \quad \text{(reversible process)}. \quad (10\text{-}1)$$

The infinitesimals may be made as small as one likes, so for all practical purposes, $p_{ext} = p_{int} \equiv p$, for short. For a reversible process, p_{int} is a well-defined quantity, which it certainly would not be if shocks or other kinds of waves were being transmitted through the system.

Exactly the same may be said for temperature in a reversible heat flow. The hotter object must have a temperature only infinitesimally above the colder, otherwise an infinitesimal change in conditions would not reverse the flow. Thus the temperature is the same (to within an infinitesimal quantity) in both the system and the surroundings if the heat interaction between them is reversible.

So, for reversible processes, two simplifications occur in the treatment of work, discussed in Sec. 5. Since there is no friction, problems of the type discussed in connection with Fig. 5-7 do not arise; the work done on a system is unambiguously the negative of the work done by the system. Also the element of work dW done on the system by the surroundings can be expressed in terms of state functions for the *system* instead of for the surroundings. Thus if only p-V work is involved, the first-law expression,

$$dE = dQ + dW = dQ - p_{ext}\, dV \quad \text{(only } p\text{-}V \text{ work)} \quad (10\text{-}2)$$

becomes

$$dE = dQ + dW = dQ - p_{int}\, dV \quad \text{(reversible process, only } p\text{-}V \text{work)} \,(10\text{-}3)$$

through the use of Eq. 10-1. Much the same happens for all contributions to dW, which for reversible processes are now expressible in terms of changes in state functions for the system. This is most convenient.

PROBLEMS

10-1. A thermodynamics book says: "Along a reversible path, all states of the system are in equilibrium with each other." Analyze the statement.

10-2. A thermodynamics book states: "The inexact differentials dQ and dW become equivalent to exact differentials under the conditions of a reversible process, or an adiabatic process, or a zero-work process." Analyze the statement.

10-3. A thermodynamics book says: "The important characteristic of a *quasistatic process* is that all state functions change so slowly that the system is at all times in a well-defined equilibrium state. Furthermore, at any time during a quasistatic change, the process can be stopped and reversed just by making an infinitesimal change in the external conditions. Quasistatic processes are commonly called reversible processes for this reason." Analyze the statement.

11 | EQUILIBRIUM AND ENTROPY*

VARIABLES DESCRIBING EQUILIBIRIUM STATES

We shall study systems at equilibrium and the processes which lead from one equilibrium state to another. The first task is to find a suitable set of variables to describe the state of an equilibrium system. The first thing that must be done is to specify what material makes up the system. This is usually done by giving the numbers of moles of each constituent compound, the mole number for the ith compound being labeled n_i. Once the material of the system has been specified, obvious initial choices for variables to describe the state are the external constraints which we can control independently from the surroundings. There is one such constraint for each kind of work that can be done on the system: the volume V for p-V work, the length of the system l for stretching, the magnetization $\mathcal{M}$ for magnetic work, the surface area $\mathcal{A}$, etc. For notational convenience, all these variables are often written as V plus the x_i, where x_1 might be the number of moles of a constituent, x_2 the magnetization, etc. If we fix the set of constraints on a particular adiabatically enclosed system, the second law says that the system will reach equilibrium. The only way we can alter that equilibrium without changing a constraint is through heat. The heat interaction changes the energy of the system but none of the constraints. The system reaches a new equilibrium after the heat interaction.

Thus for a given set of constraints, the value of the energy determines the equilibrium state for the thermodynamic system. We are neglecting cases where there is bulk motion, either translation or rotation, of the system, so

---

* Sections 11, 12, and 13 present the core of the thermodynamic principles. Unfortunately, they are also are the hardest part of the entire book. The encouraging thing about them is that there are only three such sections, that they become easier upon rereading, and that as the applications in Part II are studied, a continual cross check between Parts I and II will clarify the material in Part I.

there is no kinetic contribution to the total energy of the system. We also neglect the gravitational potential energy of the system. The energy, then, is what is called *internal energy* and is often given the symbol U. The symbol E is then retained for the total energy, including these neglected contributions (different authors use different symbols).

Out of all the myriad states the system *could* then be in with a given U, V, and x_i, it *is* in just this one. The pressure is fixed, the temperature is fixed, the densities of the various species in the various phases are fixed, everything in the equilibrium state is fixed by the energy and set of constraints. Therefore knowing the energy and constraints *should* make it possible to predict all these equilibrium properties. The means of actually calculating them is provided by the state function called *entropy*; thus its profound importance to thermodynamics.

Since U, V, and x_i specify a unique equilibrium state, it is convenient to represent these states graphically by viewing the values of U, V, and x_i as coordinates in what could be called the *space* of the thermodynamic variables. A point in this space is specified by a value for each of the variables, and with each such point is associated a unique equilibrium state of the system. If volume was the only constraint, then the space of the thermodynamic variables for that system might be two-dimensional, U and V. Since systems with two independent variables have unique mathematical properties compared to those with more, we shall think in terms of systems with one constraint x_i in addition to volume. Limitation to one x_i is done only so we do not have to visualize spaces of more than three dimensions. Only the visual picture is dependent on this limitation to one other constraint, and x_i may always be thought of as standing for all the constraints other than volume, however many there may be. Figure 11-1 shows the U-V-x_i coordinates that make up the space of the thermodynamic variables. A point is given

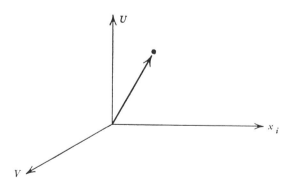

Fig. 11-1. State space with a point characterized by values of U, V, and x_i.

when values of the three coordinates are specified, and each point represents a unique equilibrium state.

KELVIN'S STATEMENT OF THE SECOND LAW

It is useful to demonstrate the following corollary of our statement of the second law: *It is impossible to have a natural process which produces no other effect than the extraction of heat from a single equilibrium reservoir and the performance of an equal amount of mechanical work.* This is another of the classical statements of the second law, due to **Kelvin*** (and Max Planck). The meaning of the statement is not always obvious at first reading. Suppose we have a single equilibrium thermal reservoir or heat bath. This could be any isolated body which has reached thermal equilibrium. It is labeled (1) in Fig. 11-2. Suppose there were a device which could violate Kelvin's

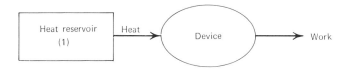

Fig. 11-2. A device (impossible) which violates Kelvin's statement of the second law.

statement. That is, it absorbs heat from the reservoir, undergoes a cycle so that it ends up in its initial state, and, in keeping with the first law, converts the energy absorbed as heat into work done on the surroundings. If such a device existed, its work could be done back on the reservoir in a manner that involved motion, and all of Fig. 11-2 could be a perpetually moving device and a violater of the second law.

Thus the scheme of Fig. 11-2 is impossible and Kelvin's statement is proved. But so far we have not said any more than could have been said in Sec. 5 about the work converters of Fig. 5-1. They too cannot really be quite 100% efficient. Neither can the converter of Fig. 11-2, which, except for the fact that its input is heat rather than work, resembles them. However, while conversion of work from one form to another can be done with efficiencies approaching 100%, conversion of heat into work is not nearly so easy, and later in this section we shall study the limitations nature imposes on this task.

---

* Sir William Thomson, later Lord Kelvin; thus the names Thomson and Kelvin are used interchangably in the scientific literature.

EXISTENCE OF THE ENTROPY

Here we prove that there exists a function of state called *entropy*, which is unchanged along any reversible adiabatic path.

The proof is based on the following corollary of Kelvin's statement of the second law: *No two states having the same constraints but different internal energies may be connected by a reversible adiabatic path.* Let us consider what might happen if this corollary were false. Then two state points, located one above the other as in Fig. 11-3, could be traversed in either

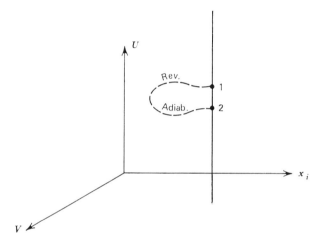

Fig. 11-3. An (impossible) reversible adiabatic path connecting two states with same V and x_i, but different U.

direction. We are not so much concerned with the direction of increasing energy, state 2 to state 1. Such a work-only path is possible, for example, through some means, as shown in Fig. 5-5a. The impossible direction is from state 1 to state 2, a work-only path to a state with lower energy and the same constraints.* The work done by the system on its surroundings in going from 1 to 2 along this path is $U_1 - U_2$. Once this was accomplished, the system could be returned to its initial state 1, simply by fixing the constraints and putting the system in thermal contact with any hot equilibrium heat reservoir. As it absorbs heat, its state point goes straight up, as indicated in Fig. 11-4, until the original state 1 is reached. The overall cyclic process

---

* We are demonstrating the content of the statement of the second law presented in 1909 by C. *Carathéodory*: *In the neighborhood of any prescribed initial state, there are states which cannot be reached by an adiabatic process.* This statement is also a corollary of our more general law of equilibrium.

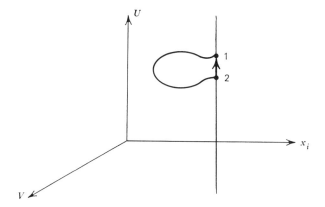

Fig. 11-4. A cyclic path which violates Kelvin's statement, possible if the reversible adiabat of Figure 11-3 were attainable.

has had one reversible and one spontaneous process, so the system could be driven with infinitesimal effort from outside. Its net effect is to absorb heat from the hot thermal reservoir and do an equal amount of work on the surroundings, thus violating Kelvin's statement. This proves the corollary.

Therefore, if one starts from state 1 and proceeds adiabatically and reversibly, then, for each set of constraints V and x_i reached, there is a single value of U that the equilibrium system may have. That is, the points accessible to a given reference state by reversible adiabatic paths form a surface in the U-V-x_i space. Such a surface is shown in Fig. 11-5; it is called a *surface of constant entropy*. If we had started from a state below or above 1 and proceeded adiabatically and reversibly from there, we would have generated a

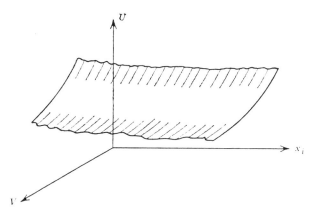

Fig. 11-5. Reversible adiabatic surface, a surface of constant entropy.

different constant entropy surface, lying below or above the one of Fig. 11-5. Nature establishes such surfaces; that is, they may be found from experiment. We can also calculate them from the first law, which for reversible processes has the form (see, for example, Eq. 10-3)

$$dU = dQ_{rev} - p\,dV + X_i\,dx_i \quad \text{(reversible process)}, \qquad (11\text{-}1)$$

and which for reversible adiabats (*adiabats* are adiabatic surfaces) is

$$dU = -p\,dV + X_i\,dx_i \quad \text{(reversible adiabats, i.e., } S = \text{const.).} \qquad (11\text{-}2)$$

We have used the conventional symbol S for entropy. Equation 11-2 prescribes the slopes* in both the V- and the x_i-directions, starting from any point and following the constant entropy surface through that point:

$$\left(\frac{\partial U}{\partial V}\right)_{S,\,x_i} = -p, \qquad \left(\frac{\partial U}{\partial x_i}\right)_{S,\,V} = X_i. \qquad (11\text{-}3)$$

CHANGES IN ENTROPY

There exists a whole family of constant entropy surfaces (sometimes called *isentropes*), each having originated from a different point on the vertical line in Fig. 11-3, and none of which can intersect lest the corollary be violated. Three such isentropic surfaces are shown in Fig. 11-6. Each of these surfaces could be described by a particular functional relation among U, V, and x_i:

$$f_j(U,\,V,\,x_i) = 0 \quad \text{(on the } j\text{th reversible adiabat).} \qquad (11\text{-}4)$$

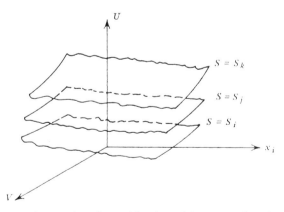

Fig. 11-6. Three isentropic surfaces with values of the entropy function assigned.

* Partial derivatives are discussed in detail in Appendix A.

It is simpler, however, to characterize the whole family by a single function of state, that is, to assign a value S_j to the entropy on the jth adiabat according to some recipe consistent with the theory. This means that the function $S = S(U, V, x_i)$ is chosen so that when S is given the constant value S_j, the result $S_j = S(U, V, x_i)$ is identical to Eq. 11-4; and similarly for other isentropes. Thus each of the isentropic surfaces of Fig. 11-6 has been labeled with the value of the entropy assigned to all points on that surface.

We now wish to find the limitations on the assignment of values for the entropy on the constant entropy surfaces. We shall find that the change in entropy dS for any infinitesimal reversible process must be given by $dS = dQ_{rev}/T(\theta)$ where T is a particular universal function of the empirical temperature only. By a **universal function** is meant one which is the same for all systems.

Once S values are assigned to all points in the U-V-x_i space, the relationship $S = S(U, V, x_i)$ may be inverted to give U as a function of the independent variables S, V, x_i:

$$U = U(S, V, x_i). \tag{11-5}$$

For arbitrary infinitesimal changes in these variables, U changes according to

$$dU = \left(\frac{\partial U}{\partial S}\right)_{V, x_i} dS + \left(\frac{\partial U}{\partial V}\right)_{S, x_i} dV + \left(\frac{\partial U}{\partial x_i}\right)_{S, V} dx_i, \tag{11-6}$$

$$dU = \left(\frac{\partial U}{\partial S}\right)_{V, x_i} dS - p\, dV + X_i\, dx_i, \tag{11-7}$$

where use was made of Eqs. 11-3. This equation must be compatible with the first law for reversible processes, Eq. 11-1:

$$dU = dQ_{rev} - p\, dV + X_i\, dx_i \qquad \text{(reversible process).} \tag{11-8}$$

Simply equating dU from Eqs. 11-7 and 11-8 yields

$$dQ_{rev} = \left(\frac{\partial U}{\partial S}\right)_{V, x_i} dS. \tag{11-9}$$

If we give the symbol T to the function $(\partial U/\partial S)_{V, x_i}$ of the state variables S, V, x_i (thus T itself is a function of state),

$$T \equiv \left(\frac{\partial U}{\partial S}\right)_{V, x_i}, \tag{11-10}$$

then Eq. 11-9 becomes

$$dS = \frac{dQ_{rev}}{T}. \tag{11-11}$$

Next we prove that T depends only on the empirical temperature θ and not on any other quantity (such as pressure, volume, density, electric field, etc.). To do this, we consider what could happen if T depended on θ and some other property, which we call Y: $T = T(\theta, Y)$. Then changing values of Y will change T, even if θ remains constant:

$$T_A = T(\theta, Y_A), \qquad T_B = T(\theta, Y_B). \qquad (11\text{-}12)$$

Let a system undergo a reversible cycle between two isentropic surfaces infinitesimally close to each other, as shown in Fig. 11-7. The system starts

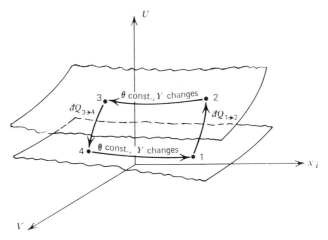

Fig. 11-7. Reversible cycle between two neighboring isentropes.

in state 1 and undergoes an arbitrary infinitesimal reversible process absorbing heat $dQ_{1\to2}$. The temperature along the entire path for this step is approximately constant, since the path is infinitesimally long and the temperature could not appreciably change in so short a path. The same is true for property Y, whose value is Y_A. The entropy change is

$$dS = \frac{1}{T_A} dQ_{1\to2} \qquad (11\text{-}13)$$

The step from state 2 to 3 is reversible and adiabatic, and therefore it is along the upper isentrope. It follows the isothermal line through point 2 (the isothermal surface cuts the isentropic surface in a line) and goes to a new value of Y, Y_B. The infinitesimal step from state 3 to 4 is reversible, absorbing just enough heat $dQ_{3\to4}$ to return the system to the original entropy it had in state 1:

$$dS = \frac{1}{T_B} dQ_{3\to4}. \qquad (11\text{-}14)$$

The last step is reversible, adiabatic, along the lower isentrope at constant temperature, back to the original state. The system has undergone a cycle, and thus its entropy must be unchanged since S is a function of state:

$$\text{Total entropy change} = \frac{1}{T_A} dQ_{1\to2} + \frac{1}{T_B} dQ_{3\to4} = 0. \tag{11-15}$$

A similar equation holds for energy:

$$\text{Total energy change} = dQ_{1\to2} + dQ_{3\to4} + dW = 0. \tag{11-16}$$

The entire cycle was conducted at constant temperature; thus it could have been done in thermal equilibrium with a heat bath at temperature θ. Since the cycle was reversible, it could be run in either direction. If the net heat absorbed were not equal to zero,

$$dQ_{1\to2} + dQ_{3\to4} = 0, \tag{11-17}$$

then dW in Eq. 11-16 could be made positive or negative simply by choosing the appropriate direction to run the cycle. A negative value for W would mean that we had violated Kelvin's statement of the second law, since the net effect of the cycle would be the extraction of heat from the reservoir at θ and the performance of an equal amount of work on the surroundings. The only way Eq. 11-17 can be compatible with Eq. 11-15 is if $T_A = T_B$, that is, if $T(\theta, Y_A) = T(\theta, Y_B)$. Thus the value of T must be independent of Y, and *it is impossible for T to depend on any property of the state of the system other than its empirical temperature.*

The fact that T is a *universal* function of θ means that once we have chosen an empirical temperature scale, *every* system must have for its T exactly the same function, $T(\theta)$. The universality of $T(\theta)$ follows from the requirement that the entropy of two subsystems viewed together as a composite single system be the sum of the entropies of the individual subsystems. This is certainly a necessary demand to make on the entropy because, in heating a composite system, either subsystem could be heated independently. Given subsystems 1 and 2, as shown in Fig. 11-8, let heat dQ be absorbed reversibly

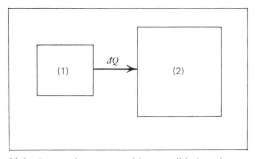

Fig. 11-8. Composite system with reversible heat interaction.

(meaning that θ_1 and θ_2 must differ only infinitesimally) by 2 from 1.   The composite is isolated from its surroundings, so the reversible process is adiabatic from the point of view of the surroundings; thus the entropy of the composite must not change.   The subsystem entropies change by $dS_1 = -dQ/T_1(\theta)$ and $dS_2 = dQ/T_2(\theta)$ respectively.   If the entropy of the composite is the sum of the entropies of its parts,

$$dS = dS_1 + dS_2 = -\frac{dQ}{T_1(\theta)} + \frac{dQ}{T_2(\theta)} = 0, \qquad (11\text{-}18)$$

$$\left[\frac{1}{T_2(\theta)} - \frac{1}{T_1(\theta)}\right] dQ = 0, \qquad (11\text{-}29)$$

$$\frac{1}{T_2(\theta)} = \frac{1}{T_1(\theta)}, \qquad T_1(\theta) = T_2(\theta). \qquad (11\text{-}20)$$

Thus T_1 must be the same function of θ as T_2 for *any* two objects; *T is a universal function of θ.*

SUMMARY

We have now established the existence of an extensive function of state, the entropy S, changes in which are given by

$$dS = \frac{dQ_{rev}}{T(\theta)} \qquad (11\text{-}21)$$

The units of T can, of course, be anything, as can those for S.   This universal function T, which plays a major role in thermodynamics, is called the **absolute temperature** or the **thermodynamic temperature**.   It would have to be found from experiment once an empirical scale θ was chosen.   It is proved in Sec. 14 that the Kelvin temperature scale may be used directly for $T(\theta)$ without destroying the validity of Eq. 11-21.   This, of course, immediately fixes the units of entropy to be energy per degree Kelvin.

Equation 11-21 shows that, if the inexact differential dQ_{rev} is divided by the absolute temperature, a perfect differential—that of the function of state entropy—is the result.   Thus one could say that *the absolute temperature is an integrating denominator for the reversible heat*, since dQ_{rev}/T is integrable in contrast to dQ_{rev}:

$$\int_1^2 dQ_{rev} = Q_{rev} \qquad \text{(describes the path between states 1 and 2),} \quad (11\text{-}22)$$

$$\int_1^2 \frac{dQ_{rev}}{T} = \int_1^2 dS = S_2 - S_1 = \Delta S \qquad \text{(independent of path, depends only on states 1 and 2).}$$

$$(11\text{-}23)$$

ENTROPY CHANGES IN IRREVERSIBLE PROCESSES

We now show that the heat absorbed by a system in *any* natural process from a source at the same temperature as the system will always be less than or equal to $T\,dS$. Consider two neighboring isentropes, as shown in Fig. 11-9, having the constant entropy values S and $S + dS$ (dS might be either

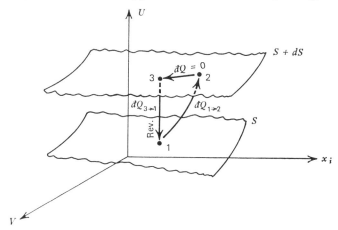

Fig. 11-9. Demonstration of Clausius' inequality.

positive or negative). Let there be some arbitrary natural process going from state 1 to state 2 with the absorption of heat $dQ_{1\to2}$ from a reservoir at the temperature of the system. Then, by a reversible adiabatic process leading from state 2 to 3, the original values of the constraints V and x_i could be recovered. The original state could then be reached by the reversible absorption of heat $dQ_{3\to1}$ from the reservoir. This infinitesimal cycle must lead to zero energy change:

$$\text{Total energy change} = dQ_{1\to2} + dQ_{3\to1} + dW = 0. \tag{11-24}$$

The net effect of the cycle is to absorb heat $dQ_{1\to2} + dQ_{3\to1}$ from the single reservoir and do an equivalent amount of work on the surroundings. Kelvin's statement of the second law says that this cannot be done, at least that positive work cannot be done on the surroundings by such a cycle. Of course, the surroundings can do work on the system at will, so we conclude that *the second law requires the net heat absorbed in this (or any) isothermal cycle not to be positive*:

$$dQ_{1\to2} + dQ_{3\to1} \leqslant 0, \tag{11-25}$$
$$dQ_{1\to2} \leqslant -dQ_{3\to1}, \tag{11-26}$$
$$dQ_{1\to2} \leqslant T\,dS_{1\to2}, \tag{11-27}$$

or

$$\boxed{dQ \leqslant T\,dS}\qquad \text{(arbitrary process).}\qquad\qquad (11\text{-}28)$$

In Eq. 11-26 the quantity $dQ_{3\to1}$ was subtracted from both sides of Eq. 11-25. Since step $3\to1$ is reversible with entropy change $-dS$, in Eq. 11-27 the $dQ_{3\to1}$ is replaced by $-T\,dS$. The resulting expression is called *Clausius' inequality*; it holds for all natural processes for which the temperature of the reservoir from which the heat is absorbed is the same as that of the system. The equal sign is applicable for reversible processes.

One might ask what happens if the reservoir and the system are not at the same temperature. In fact, the system may not even have a well-defined temperature if it is out of equilibrium. Nonequilibrium thermodynamics can say a little about this situation, but here all that can be done is to strengthen the inequality slightly. If dQ is positive and the system absorbs heat, the surroundings' temperature must be higher than the system's; the opposite is true if dQ is negative. The stronger inequality

$$dQ \leqslant T_{\text{surr}}\,dS\qquad\qquad (11\text{-}29)$$

is therefore valid for these cases.

SUMMARY AND DISCUSSION

Since an isolated system must reach equilibrium, knowledge of the energy and set of constraints is enough to fix a system's equilibrium state. There exists a function of state for equilibrium systems called the entropy S, changes in which may be found by integrating $dS = dQ_{\text{rev}}/T$ over any reversible path between the states in question. The T stands for that particular universal function of the empirical temperature called the absolute temperature, which makes dQ_{rev}/T a perfect differential. It is shown in Sec. 14 that the Kelvin scale may be used for T. For irreversible processes only the inequality $dS \geqslant dQ/T_{\text{surr}}$ may be asserted. This means that all naturally occurring *adiabatic* processes either increase the system's entropy or leave it unchanged:

$$dS \geqslant 0\qquad \text{(adiabatic process).}\qquad\qquad (11\text{-}30)$$

In particular, as an *isolated* system evolves toward its inevitable equilibrium, its entropy will increase as irreversible events occur. Calculation of the entropy change for an irreversible process cannot be made by using the dQ for that process because of the inequality of Eq. 11-29. Instead, some reversible path connecting the same initial and final states has to be imagined and then ΔS may be found by integrating dQ_{rev}/T for that path.

Although adiabatic processes can leave the entropy only larger or unchanged, if the system has a heat interaction with its surroundings, its entropy can be decreased. For example, simply losing heat reversibly to a slightly cooler reservoir decreases the entropy by $\int dQ_{rev}/T$, with dQ_{rev} being negative as the cooling occurs.

We can now write

$$dU = dQ + dW \qquad \text{(any process)}, \tag{11-31}$$

$$dU = dQ_{rev} - p\,dV + X_i\,dx_i \qquad \text{(reversible process)} \tag{11-32}$$

$$\boxed{dU = T\,dS - p\,dV + X_i\,dx_i} \qquad \begin{array}{l}\text{(any neighboring equilibrium} \\ \text{states).}\end{array} \tag{11-33}$$

In Eq. 11-32, use was made of Eq. 10-3 for reversible processes. In Eq. 11-33, substitution of $T\,dS$ was made for dQ_{rev}. The result, Eq. 11-33, is derived for reversible processes. It shows how U changes with changes in S, V, and x_i. But since U is a function of S, V, and x_i whenever these variables all have meaning, Eq. 11-33 must be correct for any process connecting two neighboring equilibrium states. The equation must depend only on the initial and final states of the system and not on whether it went from one to the other of those states by a frictionless path. The great importance of Eq. 11-33, which comes from just this fact that it relates changes in state variables independently of the particular path, will become clear in the rest of this book. Whenever quantities such as pressure and temperature are used subsequently, the system properties are tacitly assumed to be changing slowly enough for these to be well defined.

CONVERSION OF HEAT TO WORK

In Sec. 9 we noted that there were such things as heat engines, which interact with heat reservoirs in such a way as to extract thermal energy from them in the form of heat and at the same time do work on the surroundings. A sketch of a typical heat engine was given in Fig. 9-1 and is redrawn in Fig. 11-10. Suppose the engine undergoes a complete cycle, chosen so that it stops in the same state it started. It absorbs heat Q_H from a heat reservoir at temperature θ_H, heat Q_C from a reservoir at θ_C, and has work W done on it by the surroundings. The subscripts H and C refer to hot and cold respectively. If we study Fig. 11-10 as an engine and not as a refrigerator or a heat pump, then we are interested in the conditions under which W is negative, that is, when the output of the engine's cycle is work done *on* the surroundings.

We can apply the first law, Eq. 7-1, to the engine during its cycle. Since it begins and ends in the same state, its energy must be unchanged:

$$\Delta U_{\text{engine}} = Q_H + Q_C + W = 0. \tag{11-34}$$

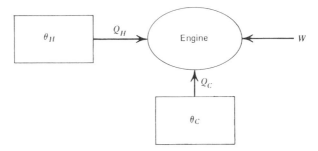

Fig. 11-10. An engine absorbing heat from two reservoirs and having work done on it by the surroundings.

The two reservoirs must be at different temperatures, otherwise they would be equivalent to a single reservoir. We showed in Sec. 9 that heat engines, just like diathermal walls, must absorb heat from the reservoir at the higher temperature and discharge it into the one at the lower. Thus if we let $\theta_H > \theta_C$, then Q_H is positive and Q_C is negative. Let us define the **efficiency** of the heat engine to be the ratio of the work, $-W$, done by the engine on the surroundings as it undergoes a cycle to the heat absorbed from the hot reservoir. Thus the efficiency is the fraction of the energy absorbed as heat which the engine turns into work. We may solve Eq. 11-34 for the efficiency:

$$\text{Efficiency} = \frac{-W}{Q_H} = \frac{Q_H + Q_C}{Q_H} = 1 + \frac{Q_C}{Q_H} \tag{11-35}$$

where Q_C is a negative quantity.

Thus *no heat engine can have* 100% *efficiency.* That we have already made clear from the second law. The best that heat engines can do is to extract heat from one equilibrium reservoir, turn some of that energy into work, and discharge the rest of the energy as heat into a cooler reservoir. If useful work is going to arise through the absorption of heat from a reservoir at θ_H, enough heat must be discharged at a low temperature θ_C to represent a sufficient approach to equilibrium or running down of that part of the universe to compensate for the work done. Creation of work takes some compensation because it could be done on *any* other body to raise its energy, no matter how hot that body already was. We shall now compute the *maximum* efficiency possible for this heat engine.

Our first consideration is how to make W as small as possible for a given change of state. We usually want to do as little work as possible to accomplish a given task, and we usually want our systems to do maximum work for us for a given change in them. This means we want W to have the smallest positive or most negative value possible. If we equate the left-hand sides of Eqs. 11-31 and 11-33, we get

$$T \, dS - dQ = dW + p \, dV - X_i \, dx_i \quad \text{(equilibrium states).} \quad (11\text{-}36)$$

Clausius' inequality, Eq. 11-29, says the left-hand side of this is greater than or equal to zero; thus the right-hand side is too:

$$dW + p \, dV - X_i \, dx_i \geqslant 0 \quad \text{(equilibrium states).} \quad (11\text{-}37)$$

Now, for a given change of state, all terms in Eq. 11-37 are fixed except for dW. Clearly, the smallest possible value for dW is obtained when the $\geqslant$ sign is, in fact, an equality, that is, when the process is reversible. Thus *the work done by the surroundings on a system to cause a given infinitesimal change of state in the system is minimized if the process is done reversibly.* Thus man constantly strives for efficiency in making his processes as nearly reversible as possible.

Thus the maximum value of the efficiency, Eq. 11-35, of the heat engine of Fig. 11-10 is achieved when the engine runs reversibly. This means that not only must there be no friction or other similar things in the engine itself but also that the part of the engine absorbing the heat from the reservoirs must be at the same temperatures as the reservoirs. For the reversible absorption of heat Q_H at the constant temperature θ_H, we can compute the entropy change in the engine:

$$\Delta S_H = \int dS = \int \frac{dQ_{\text{rev}}}{T_H} = \frac{1}{T_H} \int dQ_{\text{rev}} = \frac{Q_H}{T_H} \quad (11\text{-}38)$$

The $T(\theta_H) = T_H$ came outside the integral because θ_H is constant. Similarly, the entropy of the engine changes by Q_C/T_C during the reversible heat interaction with the cold reservoir. The work-only processes do not change the entropy because they are reversible and adiabatic. Since the engine ends up in its original state, its total entropy is unchanged, and

$$\Delta S_{\text{engine}} = \frac{Q_H}{T_H} + \frac{Q_C}{T_C} = 0. \quad (11\text{-}39)$$

Thus

$$\frac{Q_C}{Q_H} = -\frac{T_C}{T_H} \quad (11\text{-}40)$$

This result may be used in Eq. 11-35 to give the maximum efficiency of a heat engine operating between two temperatures:

$$\boxed{\text{Maximum efficiency} = -\frac{W}{Q_H} = 1 - \frac{T_C}{T_H} = \frac{T_H - T_C}{T_H}} \quad (11\text{-}41)$$

Reversible heat engines acting like this are called **Carnot engines**. They are the most efficient engines possible between the two heat reservoirs. Not only has the second law told us that heat engines must have a heat sink such as shown in Fig. 11-10 and not be like the device of Fig. 11-2, it has also let us compute the maximum fraction of the heat absorbed at the high temperature that can be turned into work, as a function of the temperatures of the source and of the heat sink.

The difficulty in converting heat to work contrasts strikingly with the theoretically *complete* conversion of one form of work to another. Briefly, the difference lies in the order or organization of the molecules or photons doing work and in the disorder or disorganization of the molecules or photons conducting heat. If heat is being conducted through a gas, the molecules are flying in all directions, bumping into each other, on the average having more random kinetic energy in the hot region than in the cold, and finally equalizing this random energy through their motion. If heat is being conducted through a solid, the molecules are rattling around randomly, bumping into each other all the time, equalizing the average random thermal energy in the process. If electromagnetic radiation is the origin of the heat interaction, then the whole spectrum of frequencies is transmitted that is characteristic of the temperature of the source of the radiation, photons going in all directions with all possible wavelengths.

On the other hand, in a work interaction enormous numbers of microscopic particles must cooperate. If a piston is being pushed back, astronomical numbers of molecules must join to some small extent in order to do their pushing in one direction at one time. If electrical work is being done, the current must flow in the desired direction, which means that vast numbers of electrons must have their motions very slightly synchronized. The same is true for other kinds of work. In fact, even electromagnetic radiation can behave like a *work* interaction when it is made monochromatic or is collimated.

Once the order is achieved and work is available, then if one is very careful to prevent friction (which is simply the destruction of the hard-won order), one can convert from one form to another with only very small losses. However, achieving the order of a work interaction out of the chaotic energy flux of a heat interaction requires an accompanying running down of part of the universe. Order involving enormous numbers of particles does not arise spontaneously from disorder, for the reasons discussed in Sec. 8. To make order out of chaos in one part of the universe, already existing order in some other part must be tapped to do the job.

One can run the engine of Fig. 11-10 backward by doing work on it. In the course of a cycle then, heat Q_C is absorbed from the cold reservoir at θ_C, and the heat Q_H at θ_H is a negative quantity. In the case of a *refrigerator*, we are interested in the maximum heat Q_C absorbed from the cold reservoir for a given amount of work W done on the refrigerator. This is found from Eq. 11-34:

$$\frac{Q_C}{W} = \frac{-Q_H - W}{W} = -\frac{Q_H}{W} - 1 \tag{11-42}$$

$$= \frac{T_H}{T_H - T_C} - 1, \tag{11-43}$$

$$\frac{Q_C}{W} = \frac{T_C}{T_H - T_C} \qquad \text{(maximum efficiency of refrigerator)}. \tag{11-44}$$

In Eq. 11-42 substitution for Q_C was made from the first law, Eq. 11-34. In Eq. 11-43 substitution for $-W/Q_H$ was made from the result, Eq. 11-41, already obtained. This result was then put over a common denominator in Eq. 11-44.

This gives a quantitative interpretation of Clausius' statement of the second law: A cycle which has no effect other than to extract heat from a cool reservoir and give it to a hot one is impossible. Refrigerators absorb heat from cool reservoirs and discharge it into warm ones. But they must be driven by work done on them by the environment, an effect which could never be undone in the environment without causing some other permanent change. Thus in order to increase the organization in part of the world by taking some thermal energy from a lower to a higher temperature, there must be an accompanying permanent effect, such as the lowering of a weight, in the rest of the world.

Finally, for the case of a *heat pump*, we run the engine backward just like a refrigerator, but we focus on $-Q_H/W$, the ratio of the amount of heat which is discharged into the high-temperature reservoir to the work required to drive the pump:

$$-\frac{Q_H}{W} = \frac{Q_C}{W} + 1, \tag{11-45}$$

$$-\frac{Q_H}{W} = \frac{T_H}{T_H - T_C} \qquad \text{(maximum efficiency of heat pump)}. \tag{11-46}$$

In Eq. 11-45 use was made of Eq. 11-42, and in Eq. 11-46 use was made of Eq. 11-43. The heat pump discharges into the reservoir being heated the heat it absorbed from the cold reservoir plus the energy represented by the work the surroundings do to force the pump through its cycle.

PROBLEMS

11-1. Humpty-Dumpty sat on a wall,
    Humpty-Dumpty had a great fall;
    All the King's horses and all the King's men
    Couldn't put Humpty together again.   [*Mother Goose*]
Discuss the truth or falseness of this famous poem in light of the second law of thermodynamics.

11-2. Calculate the maximum efficiency of a heat engine operating between the temperature of $310°C$ of superheated steam in a boiler and an ambient temperature of $20°C$.

11-3. A home is to be maintained at $22°C$, and the external temperature is $-4°C$.   The home may be heated by converting electrical work directly to heat or by using electrical work to operate a heat pump, taking thermal energy from the cold environment into the warm house.   If the heat pump acts with 100% of its theoretical efficiency, what is the ratio of costs for the two methods of heating?

11-4. A gas contained in a cylinder is in thermal contact with a large heat reservoir at temperature T.   The gas is slowly expanded at T, absorbing heat from the reservoir and doing work on its surroundings.   How is this to be reconciled with Kelvin's statement of the second law?

11-5. It has been said that one of the two small quantities, $y\,dx - x\,dy$ and $(y\,dx - x\,dy)/y^2$, is a perfect differential and the other is not.   What does that mean?

11-6. It has been said that the pressure is an integrating denominator for reversible p-V work.   What does that mean?

11-7. A thermodynamics book says: "We re-emphasize the fact that a reversible process can be defined as one which can be plotted in state space."   Analyze that statement.

11-8. The second law is often stated as "The entropy of the universe tends to a maximum."   What faults could be found with this statement?   What does it *really* mean?

11-9. A thermodynamics book says: "These findings can be summarized by the formal statement that no substance can be caused to undergo a cyclic series of operations unless at least one operation involves the absorption of heat, and at least one operation involves the rejection of heat.   This statement is clearly equivalent to the statement that all natural spontaneous changes are irreversible."   Analyze this statement.

11-10. The statement is often made by students on examinations that "the entropy of a system in a spontaneous process always increases."   Give several examples of spontaneous processes for which the system's entropy decreases.   What *can* be said about increasing entropy in these cases?

11-11. Suppose ΔS for a process is 10.00 cal/deg.   The temperature was constant at $20.00°C$.   Could Q for that process have been 3005 cal?   Why?

11-12. A freezer operates at 25% efficiency between 0 and $25°C$.   How many kcal of work are needed to remove 1 kcal of heat from the freezer?

11-13. The Pacific Gas and Electric Company, serving Northern California, generates electric power at the rate of about 15,000 Mw (as of 1970).   The steam in the generating plants is heated to about $550°C$ and is cooled in river and ocean water at about $20°C$.   What is the minimum amount of waste heat (in kcal) that must be added each day to the cooling water by PG & E?   If this heat were used to heat a square lake 1 km on a side, how deep would the lake be if its temperature rose $10°C$ per day?

12 COMPOSITION AS A VARIABLE

WHAT IS A LARGE SYSTEM?

The literature of thermodynamics is filled with references to "large systems," to the "limit of infinitely large systems," to the size of systems increasing to the "thermodynamic limit," etc. Presumably, when one talks about a *small system*, one means a system whose properties are noticeably different from those the same material would have if it were part of a large bulk, that is, a system whose properties are changed by the presence of its surfaces, edges, etc. In order to get an idea of how small a system must be to be considered "small," one needs an estimate of the thickness of the "surface region;" that is, over how great a distance do the intensive thermodynamic properties significantly change from their relatively constant values well within a body to their relatively constant values well outside it? The answer depends, of course, on the body and the kinds of molecules that make it up, but a reasonable estimate is somewhere between 5 Å and 50 Å. Thus the "volume" of the surface region of a cube of length l cm on a side will be roughly $6 \times l^2 \times (5 \text{ to } 50) \times 10^{-8}$ cm^3, or $(3 \text{ to } 30) \times 10^{-7} l^2$ cm^3. If a cube were so small that, say, 0.1% of it would lie in the surface region, its diameter would be

$$\frac{(3 \text{ to } 30) \times 10^{-7} l^2}{l^3} = 10^{-3} \tag{12-1}$$

$$l = (3 \text{ to } 30) \times 10^{-4} \text{ cm.} \tag{12-2}$$

So unless a particle's diameter is of this order or less (the size of *colloidal* particles), it will for most practical purposes be a "large system."

PARTIAL MOLAR PROPERTIES

So far we have used U, V, and x_i as a complete set of thermodynamic variables. Many times, however, it is far more convenient to control the temperature T than the internal energy U, since thermometers and constant-temperature baths are inexpensive and accurate. Also sometimes the pressure p is more convenient to control than the volume, since systems are often left open to the relatively constant pressure of the atmosphere.   For a large system of fixed composition, simply specifying T, p, and n will fix the equilibrium state if p-V work is the only kind permitted.

For large systems at fixed T, p, and composition, where n is the only variable, the values of the extensive quantities V, U, and S must be proportional to the extent of the system, that is, they must be proportional to n. This derives from the fact, previously noted, that the volume, energy, or entropy of several subsystems viewed as a composite is the sum of the volumes, energies, or entropies of the separate subsystems.   Only when surfaces and edges play a significant role does nonadditivity prevail, and these effects are negligible in large systems.   We may define the following intensive quantities: the **molar volume** v, the **molar internal energy** u, and the **molar entropy** s, all functions of T, p, and composition:

$$v = \frac{V}{n}, \qquad u = \frac{U}{n}, \qquad s = \frac{S}{n}. \tag{12-3}$$

Here we have indicated explicitly that the extensive property V is the product of the extensive mole number and the intensive molar volume, and similarly for U and V.

Now suppose we want to vary the composition by changing the numbers of moles of the various constituents present.   The independent variables are T, p, and n_i, where the notation n_i means the whole set of mole numbers. We can write the total differential of V, U, and S:

$$V = V(T, p, n_i), \qquad U = U(T, p, n_i), \qquad S = S(T, p, n_i), \tag{12-4}$$

$$dV = \left(\frac{\partial V}{\partial T}\right)_{p, n_i} dT + \left(\frac{\partial V}{\partial p}\right)_{T, n_i} dp + \sum_k \left(\frac{\partial V}{\partial n_k}\right)_{T, p, n_{j \neq k}} dn_k, \tag{12-5}$$

$$dU = \left(\frac{\partial U}{\partial T}\right)_{p, n_i} dT + \left(\frac{\partial U}{\partial p}\right)_{T, n_i} dp + \sum_k \left(\frac{\partial U}{\partial n_k}\right)_{T, p, n_{j \neq k}} dn_k, \tag{12-6}$$

$$dS = \left(\frac{\partial S}{\partial T}\right)_{p, n_i} dT + \left(\frac{\partial S}{\partial p}\right)_{T, n_i} dp + \sum_k \left(\frac{\partial S}{\partial n_k}\right)_{T, p, n_{j \neq k}} dn_k. \tag{12-7}$$

The subscript $n_{j \neq k}$ on a derivative means that all the mole numbers j are held constant except for the kth one, which is being varied.

The derivatives appearing in Eqs. 12-5 to 12-7 are properties of the system which appear over and over again in thermodynamics.   In order to integrate Eqs. 12-5 to 12-7 to find finite changes in V, U, or S, one has to know the values of these various derivatives.   In later sections, the derivatives with respect to T and p are identified in terms of easily measured quantities.   The ones important for this section are those with respect to mole numbers.   The infinitesimal change in volume, internal energy, or entropy, per fraction of a mole of substance k added, keeping T, p, and all other mole numbers constant, is called the **partial molar volume, partial molar internal energy**, or **partial molar entropy** of the kth constituent, and is given the symbol v_k, u_k, or s_k respectively:*

$$v_k \equiv \left(\frac{\partial V}{\partial n_k}\right)_{T, p, n_{j \neq k}}, \qquad u_k \equiv \left(\frac{\partial U}{\partial n_k}\right)_{T, p, n_{j \neq k}}, \qquad s_k \equiv \left(\frac{\partial S}{\partial n_k}\right)_{T, p, n_{j \neq k}}. \quad (12\text{-}8)$$

The partial molar properties are functions of T, p, and the composition in the system.   They are not properties of just the substance in question, but depend on all the other substances present.   For example, partial molar volumes are sometimes negative over certain concentration ranges.   This means that as one substance is added to a solution, the solution volume decreases.   Clearly such a property derives from the interactions among the species in the solution and not solely from the species being added.   If the system consists of a single compound, the partial molar properties of that compound are, of course, just the molar properties of the system.   If T and p are fixed in Eqs. 12-5 to 12-7, they become very simple when written in terms of the partial molar quantities:

$$dV = \sum_k v_k \, dn_k, \qquad dU = \sum_k u_k \, dn_k$$

$$dS = \sum_k s_k \, dn_k \qquad \text{(fixed } T \text{ and } p\text{)}. \qquad (12\text{-}9)$$

We now wish to integrate Eqs. 12-9 at fixed composition to find V, U, and S for large systems.   In order to do this, the fact that the composition is fixed must be noted explicitly.   Composition is often expressed by giving the value of the **mole fraction** x_k of each substance k in the system:

$$x_k \equiv \frac{n_k}{\sum_j n_j} \equiv \frac{n_k}{n} \qquad (12\text{-}10)$$

where n is the total number of moles in the system and x_k is the fraction of that total that is k.   If we fix the composition of the system, we fix all the mole fractions and just let n, the total number of moles, change.   Thus

$$n_k = x_k n, \qquad dn_k = x_k \, dn \qquad \text{(fixed composition)}, \qquad (12\text{-}11)$$

a result which may be used in Eqs. 12-9 to yield

$$dV = \sum_k v_k x_k \, dn, \qquad dU = \sum_k u_k x_k \, dn,$$

$$dS = \sum_k s_k x_k \, dn \qquad \text{(fixed } T, p, \text{ and composition)}. \qquad (12\text{-}12)$$

---

* The literature seems divided on calling these partial *molar* properties or partial *molal* properties, and some authors prefer to call them simply *partial properties*.

Now for large systems, the ratio dV/dn must be simply V/n if T, p, and composition are fixed. This is because adding a given amount of material always increases the volume by the same amount. If Eqs. 12-12 are solved for dV/dn, dU/dn, and dS/dn, the derivatives may be replaced by V/n, U/n, and S/n to yield

$$V = \sum_k v_k x_k n = \sum_k n_k v_k \qquad \text{(large system)}, \qquad (12\text{-}13)$$

$$U = \sum_k u_k x_k n = \sum_k n_k u_k \qquad \text{(large system)}, \qquad (12\text{-}14)$$

$$S = \sum_k s_k x_k n = \sum_k n_k s_k \qquad \text{(large system)}. \qquad (12\text{-}15)$$

The extensive quantities therefore are directly expressible in terms of the partial molar quantities. v_i, u_i, and s_i, which are intensive functions of T, p, and the composition.

Another approach to the derivation of Eqs. 12-13 to 12-15 is simply to integrate Eqs. 12-9 from zero material up to the final V, U, or S, noting that at constant composition the v_k, u_k, and s_k are all constants (except in the range of integration near zero, which for large systems plays a negligible role in the result). Still another approach to the integration, somewhat more elegant, is to use Euler's theorem for homogeneous functions, discussed at the end of Appendix A.

THE CHEMICAL POTENTIAL

In Sec. 11 it was deliberately left vague just what the additional variables x_i were in, say, Eq. 11-33 for the change in internal energy:

$$dU = T \, dS - p \, dV + X_i \, dx_i. \qquad (12\text{-}16)$$

Let us rule out constraints other than volume which might contribute terms to the work and restrict our additional variables simply to the set of mole numbers n_i. Then using a different notation in keeping with this added specificity, Eq. 12-16 becomes

$$dU = T \, dS - p \, dV + \sum_k \mu_k \, dn_k \qquad \text{(only } p\text{-}V \text{ work)}. \qquad (12\text{-}17)$$

The summation is over all the chemical species in the system. Writing this defines μ_k (Greek letter mu), a quantity called the **chemical potential of species k**, to be

$$\mu_k \equiv \left(\frac{\partial U}{\partial n_k} \right)_{S, V, n_{j \neq k}}. \qquad (12\text{-}18)$$

The difference between μ_k and u_k, defined in Eq. 12-8, is that in finding μ_k, the entropy and volume are kept constant; in finding u_k, the temperature and pressure are kept constant. The two may be related by "dividing Eq. 12-17 through by dn and imposing the constancy of T, p, and $n_{j \neq k}$," as discussed in Eqs. A-41 to A-44 of Appendix A:

$$\left(\frac{\partial U}{\partial n_k}\right)_{T, p, n_{j \neq k}} = T\left(\frac{\partial S}{\partial n_k}\right)_{T, p, n_{j \neq k}} - p\left(\frac{\partial V}{\partial n_k}\right)_{T, p, n_{j \neq k}} + \mu_k, \qquad (12\text{-}19)$$

and by then replacing the derivatives in this result by the symbols for partial molar quantities, defined in Eq. 12-8:

$$u_k = Ts_k - pv_k + \mu_k \qquad \text{or} \qquad \mu_k = u_k - Ts_k + pv_k. \qquad (12\text{-}20)$$

This relates μ_k to the other partial molar quantities. An intuitive feeling for the meaning of μ_k will be acquired as its properties are found and as it is put to practical use in the second part of this book. Here we can say that it has somewhat the same physical feeling as a peculiar kind of pressure variable, namely, the higher it is, the greater tendency substance k has to escape from the system.

Finally, if for each substance in the system the first form of Eq. 12-20 is multiplied by the number of moles n_k,

$$u_k n_k = Ts_k n_k - pv_k n_k + \mu_k n_k, \qquad (12\text{-}21)$$

and these equations are summed over all the species, one gets

$$\sum_k u_k n_k = T \sum_k s_k n_k - p \sum_k v_k n_k + \sum_k \mu_k n_k. \qquad (12\text{-}22)$$

With the use of Eqs. 12-13 to 12-15 for large systems, this simplifies to

$$U = TS - pV + \sum_k \mu_k n_k \qquad \text{(large system)}, \qquad (12\text{-}23)$$

a result which has occasional application. It could have been derived directly from Eq. 12-17, either by integrating from zero material at fixed T, p, and composition or by using Euler's theorem, discussed in Appendix A. Each term in Eq. 12-23 is proportional to the extent of the system. If one wanted to account for the effects on the energy of, say, the surfaces of the system, additional terms would have to be added for them.

MEASURING PARTIAL MOLAR QUANTITIES

It is easy to define partial molar quantities as we have done in terms of the values of the extensive properties of a mixture. But it is not so easy to prescribe in general how to measure extensive properties of mixtures. Of course, volumes can be measured directly, and for this reason thermodynamics books always pick the volume for their illustration. If for a given T and p one knows the volume of a solution over a range of concentrations of the constituent substances, there are a number of graphical or analytic techniques for finding the partial molar volumes of the substances as functions of the concentration of solution.

Similar techniques exist for the internal energy. In order to find U for a mixture, one can start with known amounts of the pure constituents, for which $U_i^0 = u_i^0 n_i$ are presumed known (the zero superscript means a pure substance). The solution can then be made up and if need be restored to the original T and p. If the heat and work done in this process are measured, the energy of the mixture may be written

$$U_{\text{mixture}} = \sum_k u_k^0 n_k + Q + W, \qquad (12\text{-}24)$$

and this too may be obtained over a range of concentrations.

Entropy is not so easy, however, because in order to measure entropy changes, *reversible* processes must be used. The preceding paragraph could utilize irreversible mixing to find the partial molar energy. But it is hard to add pure k reversibly to a mixture. The reverse of the process of dumping some k into a cauldron and stirring will certainly not occur if the surroundings are changed infinitesimally. Nevertheless, a reversible mixing must at least be *imaginable*, or else s_i will remain meaningless for mixtures. In the following we show such an imagined process, at least for fluids (liquids and gases). If solids are considered, it would be best to do the mixing in a fluid state and then reversibly freeze the mixture. Otherwise it is just too hard to keep the density uniform throughout solids, since molecules diffuse so slowly in them.

The *conceptual* reversible mixing process is based on the existence of a wide variety of inorganic and organic membranes which are known to have the properties of allowing certain materials to pass through, but preventing other materials from passing. By properly combining such **semipermeable membranes**, one could imagine being able to make something through which only one of the constituents in the system could pass. Then the system could be connected to a reservoir of each of its pure constituents by a membrane permeable only to that constituent. A conceptual example is shown in Fig. 12-1. The temperatures in both reservoir and system are made the same, and the pressure p_k^0 in the reservoir is adjusted to be just that which prevents passage of k either way through the membrane.

The addition of substance k to the system is carried out reversibly by infinitesimally increasing the pressure at R on the reservoir. During the addition heat can be added reversibly to the system and its volume can be

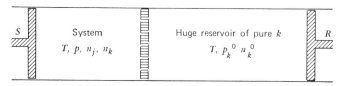

Fig. 12-1. The conceptual reversible mixing process.

changed, also reversibly, with the piston at S. For the addition of Δn_k moles of k under these circumstances, the entropy of the system is changed by an amount

$$\Delta S = s_k^0 \, \Delta n_k + \frac{Q_{\mathrm{rev}}}{T}.$$ (12-25)

Entropies of mixtures are almost never found by this hypothetical scheme, but the existence of the scheme is important conceptually.

SUGGESTED READINGS

Experimental methods of studying the extensive properties of solutions and of determining the partial molar quantities therefrom are discussed by the following:

Glasstone, Samuel, *Thermodynamics for Chemists*, D. Van Nostrand Co., Inc., Princeton, New Jersey, 1947, pp. 427–441, and references therein.

Hildebrand, J. H., **and Scott**, R. L., *Regular Solutions*, Prentice-Hall, Inc., Englewood Cliffs, N.J., 1962 (or any other book on solutions).

Lewis, G. N., **Randall**, M., **Pitzer**, K. S., **and Brewer**, L., *Thermodynamics*, McGraw Hill Book Co., New York, 2nd ed., 1961, pp. 203–213, and references therein.

Rowlinson, J. S., *Liquids and Liquid Mixtures*, Academic Press, New York, 1959, and references therein.

PROBLEM

12-1. A thermodynamics text says: "Partial molar quantities are only partially like molar properties, hence the term *partial* molar properties." Does this seem like the best explanation?

13 THERMODYNAMIC POTENTIALS AND CHANGE OF VARIABLE

INTERNAL ENERGY AS A THERMODYNAMIC POTENTIAL

We have seen in Secs. 11 and 12 that if the only kind of work we are treating is compressional, then a complete set of independent variables describing an equilibrium state is U, V, and the n_i. Therefore *every other property of an equilibrium system must be a function of U, V, and n_i.* In particular, the entropy is such a function:

$$S = S(U, V, n_i), \qquad (13\text{-}1)$$

an equation which may be solved for U to yield

$$U = U(S, V, n_i). \qquad (13\text{-}2)$$

Since U, V, and n_i suffice to specify a state and since U can be replaced by the function of S, V, and n_i, Eq. 13-2, equilibrium states might just as well be specified by S, V, and n_i as by U, V, and n_i.

If for the time being we choose S, V, n_i as our set of independent variables, then we can write the total differential of U from Eq. 13-2:

$$dU = \left(\frac{\partial U}{\partial S}\right)_{V, n_i} dS + \left(\frac{\partial U}{\partial V}\right)_{S, n_i} dV + \sum_k \left(\frac{\partial U}{\partial n_k}\right)_{S, V, n_{j \neq k}} dn_k. \qquad (13\text{-}3)$$

This may be compared with Eq. 12-17,

$$dU = T\, dS - p\, dV + \sum_k \mu_k\, dn_k. \qquad (13\text{-}4)$$

Since the left-hand sides of Eqs. 13-3 and 13-4 must be equal for arbitrary changes in the independent variables S, V, and n_i, they must be equal term by term (for proof, see Eq. A-12 in Appendix A):

$$\left(\frac{\partial U}{\partial S}\right)_{V, n_i} = T, \qquad (13\text{-}5)$$

$$\left(\frac{\partial U}{\partial V}\right)_{S, n_i} = -p, \qquad (13\text{-}6)$$

$$\left(\frac{\partial U}{\partial n_k}\right)_{S,\,V,\,n_{j \neq k}} = \mu_k. \tag{13-7}$$

These results are all familiar (see Eqs. 11-10, 11-3, and 12-18). The left-hand side of Eq. 13-5 is a function of the variables S, V, and n_i, obtained from Eq. 13-2 by differentiation. The right-hand side, T, equals this function, so if $U(S, V, n_i)$ is known, the relationship among T, S, V, n_i is immediately obtainable. It is simply Eq. 13-5. Similarly, Eq. 13-6 gives the relationship among p, S, V, n_i, and Eq. 13-7 gives it among μ_k, S, V, n_j.

By eliminating variables between various pairs of these equations, and then by subsequent eliminations among the resulting equations, expressions may be found relating *any* set of four (or more, depending on the number of constituents) of the variables S, U, V, n_i, T, p, μ_i. This is because there are only three independent variables (for single-component systems—for every additional chemical constituent there is another mole number), so every set of four must be implicitly related. For example, the so-called *thermal equation of state* is the relation coupling T, p, V, n_i:

$$f(T, p, V, n_i) = 0. \tag{13-8}$$

An example of such an equation is the ideal-gas law, $pV - nRT = 0$. The thermal equation of state for a substance may be obtained from knowing $U(S, V, n_i)$ for that substance in the following way: Write down Eqs. 13-2, 13-5, and 13-6, the latter two found from the first by differentiation. Then from these three equations eliminate the undesired quantities S and U by algebraic substitution. The result is the desired relationship coupling T, p, V, n_i. By similar means, and of course by interpreting the resulting equations physically, full thermodynamic information can be obtained about a system, all from knowing the single function, $U(S, V, n_i)$. The relation $U = U(S, V, n_i)$ is called a **fundamental relation**, and the internal energy U, when viewed as a function of S, V, and n_i, is called a **thermodynamic potential** because so many important quantities may be derived from it by differentiation.

CHANGE OF VARIABLE

Clearly it would be nice to know $U = U(S, V, n_i)$ for various kinds of materials and thus have full predictive ability. Unfortunately we never do. One reason is that S, V, and n_i are not the most convenient independent variables. Use of entropy as an independent variable in experiments is impractical in the extreme. There are no simple meters for measuring entropy and no simple way to assure that a system has constant entropy in a process. Entropy measurements commonly involve careful calorimetry in the measuring of the heat for various processes, and calorimetric experiments are notoriously difficult to perform. Even adiabatic conditions are hard to achieve in the laboratory. On the other hand, precise thermometers are available for measuring temperature, and thermostats can control temperature with great precision. Thus T is more practical as a choice of independent variable than S (or U, for that matter). Furthermore p is sometimes preferable to V as an independent variable, and sometimes μ_i is preferable to n_i. Thus it is important to find the quantity that functions as a thermodynamic potential for each appropriate choice of independent variables.

A careful discussion of this problem is given in Appendix B. Here we simply accept the answer found there: We define, quite arbitrarily, a new function A:

$$A \equiv U - TS. \qquad (13\text{-}9)$$

Its total differential may be written

$$dA = dU - T\,dS - S\,dT, \qquad (13\text{-}10)$$

and dU may be substituted from Eq. 13-4 into Eq. 13-10 to obtain

$$dA = -S\,dT - p\,dV + \sum_k \mu_k\,dn_k. \qquad (13\text{-}11)$$

From this we see that if only we knew

$$A = A(T, V, n_i), \qquad (13\text{-}12)$$

then all the quantities S, p, μ_i could be found simply by differentiation:

$$S = -\left(\frac{\partial A}{\partial T}\right)_{V, n_i}, \qquad p = -\left(\frac{\partial A}{\partial V}\right)_{T, n_i}, \qquad \mu_k = \left(\frac{\partial A}{\partial n_k}\right)_{T, V, n_{j \neq k}} \qquad (13\text{-}13)$$

The function $A(T, V, n_i)$ thus serves as a thermodynamic potential for the independent variables T, V, n_i. It is $A = U - TS$ that we should find if we want full predictive ability for those variables. This function is called the *Helmholtz free energy*, and is sometimes given symbols other than A.

Thus each set of independent variables has its own thermodynamic potential. Appropriate potentials are called **Legendre transforms** of the internal energy and are discussed in Appendix B. Commonly used Legendre transforms of the energy, shown as functions of the variables appropriate for them, are defined as follows:

$$U(S, V, n_i), \tag{13-14}$$

$$H(S, p, n_i) \equiv U + pV, \tag{13-15}$$

$$A(T, V, n_i) \equiv U - TS, \tag{13-16}$$

$$G(T, p, n_i) \equiv U - TS + pV, \tag{13-17}$$

$$\Psi(T, V, \mu_i) \equiv U - TS - \sum_i \mu_i n_i, \tag{13-18}$$

$$\Omega(T, p, \mu_i) \equiv U - TS + pV - \sum_i \mu_i n_i. \tag{13-19}$$

The names of these functions are the following: U is the **internal energy**; H is the **enthalpy** (by some authors the "heat content," a poor expression, pedagogically); A is the **Helmholtz free energy** (by physicists, sometimes, simply the "free energy"); G is the **Gibbs free energy** (by chemists, sometimes, simply the "free energy" and by physicists, sometimes, the "free enthalpy"); Ψ is the **grand potential function** (or "grand canonical potential function"); and Ω might be called the **null function**, but no name has been settled on for it. Notation is not standard for any of these functions; different authors use different symbols for the same potential, and different authors use the same symbol for different potentials. Thus one must always check to see how the author has defined his symbols. It is especially maddening when an author simply introduces a symbol, calls it the "free energy," and never defines it further!

Although it might seem hard to remember the appropriate correction terms to U in Eqs. 13-14 to 13-19, the job becomes easy when one thinks about the equations for the total differentials of the various Legendre transforms:

$$dU = T\,dS - p\,dV + \sum_i \mu_i\,dn_i, \tag{13-20}$$

$$dH = T\,dS + V\,dp + \sum_i \mu_i\,dn_i, \tag{13-21}$$

$$dA = -S\,dT - p\,dV + \sum_i \mu_i\,dn_i, \tag{13-22}$$

$$dG = -S\,dT + V\,dp + \sum_i \mu_i\,dn_i, \tag{13-23}$$

$$d\Psi = -S\,dT - p\,dV - \sum_i n_i\,d\mu_i, \tag{13-24}$$

$$d\Omega = -S\,dT + V\,dp - \sum_i n_i\,d\mu_i. \tag{13-25}$$

All these equations are derived immediately (see Prob. 13-1) from Eq. 13-20 and the definition of the appropriate function.   By deriving all of these, the reader will see the scheme of things.   Additional motivation is provided in Appendix B.   The set of Eqs. 13-20 to 13-25 are sometimes called the **Gibbs equations**.

At this point we could write down the 18 equations expressing the derivatives of these six functions; examples are Eqs. 13-5 to 13-7 which follow from Eq. 13-20, and Eqs. 13-13 which follow from Eq. 13-22.   However, the reader could write these down for himself more profitably.

LARGE SYSTEMS

It is interesting to consider what the various thermodynamic potentials are for large systems, starting with the internal energy, which was found in Eq. 12-23.   Expressions for the others follow (see Prob. 13-2) immediately from their definitions:

$$U = TS - pV + \sum_i \mu_i n_i \quad \text{(large system)},\tag{13-26}$$

$$H = TS + \sum_i \mu_i n_i \quad \text{(large system)},\tag{13-27}$$

$$A = -pV + \sum_i \mu_i n_i \quad \text{(large system)},\tag{13-28}$$

$$G = \sum_i \mu_i n_i \quad \text{(large system)},\tag{13-29}$$

$$\Psi = -pV \quad \text{(large system)},\tag{13-30}$$

$$\Omega = 0 \quad \text{(large system)}.\tag{13-31}$$

Equation 13-29 shows that *the chemical potential is the partial molar Gibbs free energy*.   We already were aware of this, however, since it is shown by Eq. 12-20, if that equation is viewed in light of the definition of Gibbs free energy, Eq. 13-17.   We could use the symbol g_i rather than μ_i to emphasize this equivalence, but history and convention dictate otherwise.

The fact that Ω is zero for large systems means that it is a measure of the *smallness* of a system.   Any corrections to Eq. 12-23 due to surfaces or edges are the only things left to comprise Ω.   If $\Omega(T, p, \mu_i) = 0$, this is a functional relation among the set of intensive variables.   Thus if all but one are given, that one is fixed.   The reason there is one less degree of freedom among the intensive quantities is that the system could be any size, contain any total amount of material, and still have the same values of T, p, and μ_i.   Therefore there must be one less independent intensive variable than total independent variables, since one independent variable must be extensive and thus reflect the total system size. Some authors use a terminology which says that the set of intensive variables actually specifies the *state* of a system, in which case all an extensive quantity does is tell *how much* of the system in that state there is.

Since the intensive quantities are related, changes in them are coupled by Eq. 13-25, which equals zero for large systems (if Ω is the constant zero, $d\Omega$ must equal zero too):

$$0 = -S\,dT + V\,dp - \sum_k n_k\,d\mu_k \qquad \text{(large system)}. \qquad (13\text{-}32)$$

This relation among the intensive variables is called the **Gibbs-Duhem equation**.

Other thermodynamic potentials based on U and a complete set of Legendre transforms starting with the entropy instead of the energy are capable of definition and use. After becoming familiar with the ones given here, especially with U, H, A, and G, one should have no trouble using new ones. A disadvantage in having to use transforms at all is the loss of physical intuition for them. They do, however, have a certain energylike feeling, which will be acquired as one sees their specific properties and uses them in a variety of problems. Despite their physical obscurity, they are what give thermodynamics is great utility. Without the freedom to use whatever variables are convenient, the subject would have only academic interest, at most.

THE MAXWELL RELATIONS

The fact that thermodynamic variables are all first derivatives of a potential means that derivatives of the variables themselves are coupled by the equality of mixed second derivatives. This is discussed in Appendix A, Eq. A-24. For example, using the Helmholtz function, one knows

$$\left(\frac{\partial}{\partial V}\right)_T \left(\frac{\partial A}{\partial T}\right)_V = \left(\frac{\partial}{\partial T}\right)_V \left(\frac{\partial A}{\partial V}\right)_T. \qquad (13\text{-}33)$$

Now, if $-S$ is substituted for $(\partial A/\partial T)_V$ and $-p$ for $(\partial A/\partial V)_T$ in Eq. 13-33 (identifications already obtained in Eq. 13-13), the result is

$$\left(\frac{\partial S}{\partial V}\right)_{T,\,n_i} = \left(\frac{\partial p}{\partial T}\right)_{V,\,n_i}. \qquad (13\text{-}34)$$

Both the left- and right-hand sides of Eq. 13-34 are functions of T, V, and n_i. They are the same functions, regardless of their apparently different origins. This is one of an enormous number of such expressions based on the Gibbs relations and the equality of the mixed second derivatives. They are called the **Maxwell relations** and some of them are very useful indeed. For example, Eq. 13-34 is important, since it is often very difficult to find the left-hand side experimentally. However, the right-hand side is known simply from the thermal equation of state, and p-V-T measurements are relatively easy to make and are very well tabulated for common substances. Here is an excellent example of using thermodynamic theory to relate something hard to find to something else that is easier.

In addition to Eq. 13-34 which came from the Helmholtz function, the Gibbs function gives the similar very useful Maxwell relation:

$$\left(\frac{\partial S}{\partial p}\right)_{T, n_i} = -\left(\frac{\partial V}{\partial T}\right)_{p, n_i}, \qquad (13\text{-}35)$$

which the reader may verify (see Prob. 13-3). These may be the only two Maxwell relations one will ever need, but they are used very often. Like most thermodynamic expressions, these are best not memorized, but derived when needed from the Gibbs relations.

GIBBS FREE ENERGY AND NONCOMPRESSIONAL WORK

The first law for a closed system (chemical reactions might be occurring within the system) is $dU = dQ + dW$. For reversible processes this becomes

$$dU = T\,dS - p\,dV + dW_{other} \qquad \text{(closed system, reversible process)}, \quad (13\text{-}36)$$

where work other than compressional has been noted separately. In Eq. 5-13 a host of different contributions to this term is given. Now, if $G \equiv U - TS + pV$ is differentiated and dU substituted from Eq. 13-36, the result is

$$dG = -S\,dT + V\,dp + dW_{other} \qquad \text{(closed system, reversible process)}.$$

$$(13\text{-}37)$$

At constant T and p this becomes

$$dG = dW_{other} \qquad \text{(closed system, reversible process, } T \text{ and } p \text{ constant)}.$$

$$(13\text{-}38)$$

The change in Gibbs free energy for a closed system undergoing a reversible process at constant temperature and pressure equals the work other than compressional done on the system during the process.

Thus if only p-V work is done during a reversible process at constant T and p, the Gibbs free energy is unchanged. An example of such a process is a simple phase transition, such as the boiling of water at its boiling point under a fixed pressure. Such a phase transition must therefore not change the value of G. This will prove to be a most useful result; it is extended somewhat in Sec. 15 and is applied extensively in Secs. 25 through 30.

PROBLEMS

13-1. Derive Eqs. 13-21 through 13-25 from the Gibbs equation for internal energy, Eq. 13-20, and the definitions of the various thermodynamic potentials.

13-2. Prove Eqs. 13-27 through 13-31 from Eq. 12-23 and the definitions of the appropriate functions.

13-3. Derive Eq. 13-35.

13-4. Starting from the fundamental representation given in Appendix B for a monatomic gas, Eq. B-9, calculate T, p, and μ as functions of S, V, and n.

13-5. Starting from Eq. B-9, verify the ideal-gas law, Eq. B-10.

13-6. Derive Eq. B-11 from Eq. B-9.

13-7. Derive Eq. B-21 from Eq. B-9.

13-8. Starting from the fundamental representation of Eq. B-21, find S, p, and μ as functions of T, V, and n. Compare the expression for p with Eq. B-10.

13-9. Derive Eq. B-11 from Eq. B-21.

13-10. One sometimes sees the surface tension γ (see Sec. 5) defined as the Gibbs free energy per unit surface. Why is this definition justified?

14 EQUALITY OF THERMODYNAMIC AND KELVIN TEMPERATURES

We have so far been using the symbol T to refer to the thermodynamic temperature, that is, that particular function of an empirical temperature θ which makes $dQ_{rev}/T(\theta)$ a perfect differential. However, we still have not determined how to measure the thermodynamic temperature, that is, how to find out what this particular function of the empirical temperature is. In this section, it is shown that the Kelvin temperature scale discussed in Sec. 9 is a satisfactory choice for the absolute temperature. We find that use of θ_K, defined from real-gas behavior in the limit of vanishing density, as T requires for consistency with the second law a further property of real gases: in the limit of vanishing density, their internal energies must become functions of the temperature only. Since this is an observed property of real gases, the Kelvin scale is indeed satisfactory for use as the absolute thermo-dynamic temperature.

The absolute temperature scale enters thermodynamics as the integrating denominator of the reversible heat, something which is fully expressed in the Gibbs relation for a closed system,

$$dU = T\, dS - p\, dV \qquad (n_i \text{ constant}). \qquad (14\text{-}1)$$

A mathematical consequence of this ("division by dV and imposition of constant T," see Eqs. A-41 to A-44) is

$$\left(\frac{\partial U}{\partial V}\right)_T = T\left(\frac{\partial S}{\partial V}\right)_T - p \qquad (n_i \text{ constant}). \qquad (14\text{-}2)$$

The Maxwell relation from the Helmholtz function, Eq. 13-34, may be used to replace the $(\partial S/\partial V)_T$ by $(\partial p/\partial V)_V$. This is sometimes called *Maxwellizing* the derivative.

$$\left(\frac{\partial U}{\partial V}\right)_T = T\left(\frac{\partial p}{\partial T}\right)_V - p \qquad (n_i \text{ constant}). \qquad (14\text{-}3)$$

Now we can divide through by p to yield

$$\frac{1}{p}\left(\frac{\partial U}{\partial V}\right)_T = \frac{T}{p}\left(\frac{\partial p}{\partial T}\right)_V - 1 \qquad (n_i \text{ constant}). \qquad (14\text{-}4)$$

This result, where T is the absolute temperature, is implied by our formal thermodynamics.

On the other hand, the Kelvin or "ideal-gas" temperature scale is based on experimental properties of real gases in the limit of vanishing density, assigning numbers to temperatures through the expression, Eq. 9-4,

$$\theta_K = \frac{1}{R} \lim_{\rho \to 0} \frac{pV}{n} \qquad (\text{any real gas}), \qquad (14\text{-}5)$$

where R is 0.08205967 l-atm/deg-mole. The question is whether θ_K may be used for T. In the limit of vanishing density, the pressure for real gases is, from Eq. 14-5,

$$p = \frac{nR\theta_K}{V} \qquad (\text{vanishing density}). \qquad (14\text{-}6)$$

Thus

$$\left(\frac{\partial p}{\partial \theta_K}\right)_V = \frac{nR}{V} \qquad (\text{vanishing density}). \qquad (14\text{-}7)$$

If we use θ_K for T, the right-hand side of Eq. 14-4 assumes the value zero in the limit of vanishing density:

$$\left(\frac{\theta_K}{p}\right)\left(\frac{nR}{V}\right) - 1 = 0 \qquad \text{(vanishing density)}. \qquad (14\text{-}8)$$

Therefore the use of the Kelvin temperature scale θ_K for the absolute thermodynamic temperature T can be made consistent with the demands of formal thermodynamics only if all real gases show that the left-hand side of Eq. 14-4 approaches zero:*

$$\lim_{\rho \to 0} \frac{1}{p}\left(\frac{\partial U}{\partial V}\right)_\theta = 0 \qquad \text{(all gases)}. \qquad (14\text{-}9)$$

Happily, when experiments are performed on real gases in the limit of vanishing density, not only is the property, Eq. 9-2, observed,

$$\lim_{\rho \to 0} \frac{pV}{n} = f(\theta \text{ only}), \qquad (14\text{-}10)$$

where f is a universal function of the empirical temperature only, but also the internal energy of real gases becomes a function of temperature only at low densities, and Eq. 14-9 is verified experimentally:

$$\lim_{\rho \to 0} U = nu(\theta \text{ only}). \qquad (14\text{-}11)$$

The internal energy of 1 g of oxygen in a 100-l vessel at 20°C is almost exactly the same as that it would have in a 200-l vessel at 20°C. The density is halved, the volume is doubled, the pressure is halved, but if the temperature is not changed, the internal energy remains constant. Only by changing to a new temperature is the internal energy changed for a dilute real gas. This is further discussed in Sec. 23. It is, in fact, a reasonable approximation for most real gases at pressures up to a couple of atmospheres.

Thus the Kelvin scale, while it may not be the simplest thermodynamic temperature scale, at least is a permissible one. In most scientific writings, when T appears in an equation treating specific numerical values of the temperature, the Kelvin scale is implied. This is usually made apparent by the appearance of the gas constant R multiplying T where it appears in these equations.

---

* The right-hand side of Eq. 14-3 also vanishes if θ_K is used for T. The reason why Eq. 14-4 is used instead is that Eq. 14-9 is a stronger demand on the behavior of real gases than

$$\lim_{\rho \to 0}\left(\frac{\partial U}{\partial V}\right)_\theta = 0,$$

since p goes to zero as ρ goes to zero. Thus Eq. 14-9 requires that $(\partial U/\partial V)_\theta$ go to zero *faster than p.*

PROBLEMS

14-1. Given the experimental results discussed in this section, what can be said about the enthalpy of gases in the low-density limit?

14-2. A thermodynamics text states: "The fact that $U = U(T$ only) and $H = H$ (T only) is an equivalent criterion for an ideal gas to the fact that $pV = nRT$." Analyze this statement.

15   CRITERIA FOR EQUILIBRIUM

This section uses the first and second laws of thermodynamics to find the conditions governing systems at equilibrium. The results we find are that *all parts of an equilibrium system (a) which are in thermal contact must be at the same temperature; (b) which can exchange volume without friction must be at the same pressure, (c) which can exchange matter of species i through diffusion must have the same chemical potential for species i.*

EQUALITY OF TEMPERATURES

Consider two subsystems, 1 and 2, connected by a diathermal wall, but the composite isolated from the surroundings. Their volumes remain fixed, but suppose that, as a result of their interaction through the wall, dQ_1 is absorbed by 1 from 2 and dQ_2 is absorbed by 2 from 1. We have already seen in Prob. 7-1 that $dQ_1 = -dQ_2$, a fact that follows from applying the first law to the isolated composite, which, of course, keeps a constant internal energy:

$$dU = 0 = dQ_1 + dQ_2. \tag{15-1}$$

A sketch of these interacting subsystems is given in Fig. 15-1. The overall isolated system will evolve spontaneously in keeping with Clausius' inequality through additional heat interaction so long as the entropy can increase thereby.

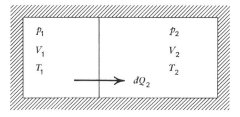

Fig. 15-1. Subsystems interacting through diathermal wall.

Therefore we shall compute the entropy change for the two subsystems caused by their heat interaction. To do this we use the Gibbs equation (Eq. 13-20) for closed subsystems:

$$dU_1 = T_1\,dS_1 - p_1\,dV_1, \qquad dU_2 = T_2\,dS_2 - p_2\,dV_2. \qquad (15\text{-}2)$$

Since both dV_1 and dV_2 are zero, we can find the total entropy change:

$$dS = dS_1 + dS_2 = \frac{1}{T_1}\,dU_1 + \frac{1}{T_2}\,dU_2, \qquad (15\text{-}3)$$

$$dS = \frac{1}{T_1}\,dQ_1 + \frac{1}{T_2}\,dQ_2 \qquad (15\text{-}4)$$

$$dS = dQ_2\left(\frac{1}{T_2} - \frac{1}{T_1}\right) = dQ_2\,\frac{T_1 - T_2}{T_1 T_2}. \qquad (15\text{-}5)$$

In Eq. 15-3 substitution was made from Eq. 15-2; in Eq. 15-4, recognition was made of the fact that the subsystems changed their energy only through their heat interaction; in Eq. 15-5, $dQ_1 = -dQ_2$ was substituted by using Eq. 15-1.

Clausius' inequality, Eq. 11-30, states that adiabatic processes where $dQ = 0$ must either increase the entropy or leave it unchanged. The mechanism whereby this is accomplished is through the infinitesimal fluctuations that are always occurring naturally within any system. These fluctuations will test out all the states near to whatever one the system is in. The system will be unable to get into those which significantly decrease the entropy, but nothing prevents its going into nearby states which increase the entropy. Once the system is in one of these states, however, it cannot go back, since the reverse process would then represent an entropy decrease. By this procedure the isolated system evolves to a final equilibrium state, in which all neighboring states lie at lower or the same entropy.

Thus heat will flow between the two subsystems whenever Eq. 15-5 is positive. When $T_1 > T_2$, dQ_2 is positive; thus heat flows from 1 to 2. When $T_1 < T_2$, dQ_2 must be negative; thus heat flows from 2 to 1. *Energy flows in the form of heat from regions of high temperature to those of low temperature until the temperatures become the same.* Of course, we carefully built this result into the structure of thermodynamics from the beginning, so it is not surprising. Still, it is gratifying to conclude something that is consistent with what we know must be true.

EQUALITY OF PRESSURES

Consider two subsystems, this time able to exchange volume by the expansion of one at the expense of the other. An example is sketched in Fig. 15-2,

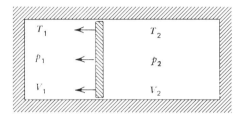

Fig. 15-2. Subsystems interchanging volume via a movable adiabatic wall.

chosen so that, as the subsystem volumes change, the area of the surface between them remains unchanged. Otherwise the area must be treated as a variable, a complication which is not treated in this book.

An infinitesimal motion of the wall changes the entropy of each subsystem by an amount obtained from Eq. 15-2:

$$dS = dS_1 + dS_2 = \frac{1}{T_1} dU_1 + \frac{p_1}{T_1} dV_1 + \frac{1}{T_2} dU_2 + \frac{p_2}{T_2} dV_2. \quad (15\text{-}6)$$

The total volume $V = V_1 + V_2$ remains unchanged, so the volume lost by one subsystem is gained by the other:

$$0 = dV_1 + dV_2. \quad (15\text{-}7)$$

Since no heat is absorbed by either subsystem, the first law may be used to find the energy changes:

$$dU_1 = dW_1 = -p_2 \, dV_1, \qquad dU_2 = dW_2 = -p_1 \, dV_2. \quad (15\text{-}8)$$

For this irreversible process, we must be careful to use the work done *on* each subsystem, which is, of course, the same as the work done *by* the other subsystem, since the only interaction of each subsystem is with the other. We note that despite conservation of total volume and total energy, the total *internal* energy is not necessarily conserved if $p_1 \neq p_2$, because some *kinetic* energy may be imparted to the movable wall or to parts of the system.   Use of Eqs. 15-7 and 15-8 in Eq. 15-6 yields

$$dS = \left(\frac{p_2}{T_1} - \frac{p_1}{T_1} - \frac{p_1}{T_2} + \frac{p_2}{T_2} \right) dV_2, \tag{15-9}$$

$$dS = \left(\frac{1}{T_1} + \frac{1}{T_2} \right) (p_2 - p_1) \, dV_2. \tag{15-10}$$

The wall separating the subsystems will move in such a way that dS is positive whenever $p_2 \neq p_1$.   When $p_2 > p_1$, dV_2 will be positive, and vice versa. Thus, one might say that *volume "flows" from regions of low pressure to regions of high pressure*.   At equilibrium the pressures will be equal.   Of course, if there is friction in a piston between the subsystems, they may be prevented from trying out configurations in which their entropy is higher; the pressures will necessarily be equal only where friction plays no role. Again, the result we have found is familiar from our everyday experience, namely that bulk flows of matter occur, where possible, from regions of high pressure to regions of low pressure.

EQUALITY OF CHEMICAL POTENTIALS

Consider two subsystems, this time able to exchange matter of species j through some kind of boundary, say, a membrane permeable only to j. Let the temperatures of the subsystems be the same, let their volumes be fixed, and let $dn_{j,2}$ moles of j pass from subsystem 1 to 2.   This is sketched in Fig. 15-3.   From the Gibbs equation for the internal energy,

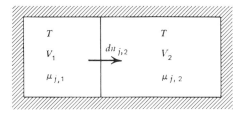

Fig. 15-3. Subsystems interchanging matter via a semipermeable membrane.

$$dU = T \, dS - p \, dV + \sum_i \mu_i \, dn_i, \tag{15-11}$$

separately valid for each subsystem, we get

$$dS = dS_1 + dS_2 = \frac{1}{T} dU_1 - \frac{\mu_{j,1}}{T} dn_{j,1} + \frac{1}{T} dU_2 - \frac{\mu_{j,2}}{T} dn_{j,2} \tag{15-12}$$

Conservation of energy and of matter,

$$dU_1 + dU_2 = 0, \qquad dn_{j,1} + dn_{j,2} = 0, \qquad (15\text{-}13)$$

reduces Eq. 15-12 to simply

$$dS = \left(\frac{\mu_{j,1} - \mu_{j,2}}{T}\right) dn_{j,2}. \qquad (15\text{-}14)$$

Material of type j will flow from subsystem 1 to 2 as long as the entropy increases thereby, that is, whenever $\mu_{j,1} > \mu_{j,2}$. It will flow from 2 to 1 when $\mu_{j,1} < \mu_{j,2}$. Thus at constant temperature, *matter flows from regions of high chemical potential to regions of low chemical potential, until at equilibrium μ_j is the same in all regions of the system into which it is possible for j to get.* This conclusion proves very useful in a host of applications.

This result has been anticipated in the discussion at the end of Sec. 13, where it was noted that reversible phase transitions occur at constant Gibbs free energy; thus the molar Gibbs free energy μ must be the same in two phases which are coexisting at the same T and p. Nevertheless, the more general result obtained here may not seem intuitively obvious, and a brief molecular digression may be in order.

Why is it that molecules move to regions of lower $h - Ts$ rather than simply to regions of lower h (no serious error arises if one views enthalpy and energy as similar in this discussion)? Our everyday experience suggests that energy predicts the directions of processes. Since billiard balls and electric charges seek positions of lower energy, certainly molecules do too. Sodium hydroxide dissolves in water rather than remaining insoluble, and the solution is clearly a state of lower energy for the NaOH, since great amounts of heat are given off. Steam condenses at 100°C and the heat given off is a measure of the lower energy the molecules have in the liquid, where they are all attracting each other, as compared to the vapor, where they are a long way apart. On the other hand, NH_4NO_3 also dissolves in water, but when it does the solution cools down, so this spontaneous process leaves the NH_4NO_3 in a state of higher energy. Similarly liquid water heated above 100°C changes to vapor; the H_2O molecules spontaneously go to an environment of decidedly higher energy.

Of course, molecules, like billiard balls, are attracted to regions where their energies are low. The force of attraction is the slope of the plot of energy versus distance, just like it is for billiard balls. But interatomic forces are weak compared to the thermal energies of the Brownian motion-type randomly acquired by the molecules from their neighbors. For them, the advantage of lower energy can be overcome by something else, namely, the advantage of higher entropy. In the discussion of Clausius' inequality, Eq. 11-28, we noted that for isolated systems, for which all states have the same energy, the entropy alone determines the equilibrium condition.   What, physically, is this entropy that helps govern equilibria?

Suppose one had a room filled with frogs which were hopping about.   Let part of the floor be painted blue and the rest be painted yellow.   The relative populations of frogs in the two regions would depend solely on their relative areas.   There is an analogy between the random jumping of the frogs and the random Brownian-type motion of the molecules.   The areas of the two regions on the floor are analogous to the entropies of two states for molecules; for example, the molecules in a crystal are constrained not only to sit right next to their neighbors but also they most almost never change neighbors and must keep their rows neat.   Solids have low entropies.   In liquids the molecules are still constrained to sit next to their neighbors, but with very little effort they can slide over to some new neighbors, and all attempt to maintain neat rows is abandoned.   Liquids have higher entropies than crystals.   In gases, molecules are free to fly all through the container and bang into whatever they please.   Gases have the highest entropies, in consequence.   Thus it is that entropy is a measure of disorder, of randomness, of " space " or " area " available to the molecules in a particular state, quite analogous to the area painted one color on our hypothetical floor.   The painted floor with hopping frogs is not unlike an isolated system and its state of maximum entropy.

Now suppose we elevate the yellow parts of the floor above the blue.   The equilibrium density of frogs will be lower in the high-energy yellow region and higher in the blue.   This is because the frogs are jumping with a distribution of heights over a range from small to large.   Thus only a fraction of the jumps a given frog might make could conceivably take him up from blue to yellow; the rest of his jumps are not high enough.   However, any jump is enough to get a frog off the edge of a yellow region and down into the blue. The greater the energy difference between the two states, the more favored is the one with lower energy.   Entropy still plays its role, but a less important one.

Temperature plays an important role in this picture.   As the temperature is increased, the frogs become more active.   They not only jump more often but higher jumps become more favored.   Great heights, which are almost impossible at low temperatures, become common, and the little jumps characteristic of low temperatures become relatively rare.   Thus as T increases, the frogs become less aware of the energy difference between the states because their average jump is more than enough to clear the barrier.   Therefore, at high temperatures, molecules seek the state with high entropy.   Similarly, at low temperatures, the frogs have trouble clearing even the smallest energy barrier, so when T is small a low value of enthalpy is favored.   Thus it is not at all unreasonable that a quantity like $h - Ts$ emerges as the driving force for the diffusive flow of matter.

SYSTEMS WHICH ARE NOT ISOLATED

Most systems one encounters are not isolated from their surroundings. Rather than having fixed energies, such systems might have fixed temperatures determined by their being in thermal contact with thermal reservoirs in the surroundings.   Rather than having fixed volumes, such systems might have their pressures fixed, perhaps by being open to the ambient pressure of the atmosphere.   Despite this, the criteria for equilibria found in this section must apply; indeed, they apply to all systems, whether isolated or not.   The reason was mentioned in Sec. 8, where it was noted that equilibrium systems that are not isolated from their surroundings must still be capable of being approximated—immediate surroundings plus system—by an isolated equilibrium supersystem.   Thus how a system gets to its equilibrium condition is unimportant.   Whether temperature or energy was controlled from without is immaterial; at equilibrium all parts of the system and the surroundings will have the same temperature if they are in thermal contact.   The same holds for pressure and chemical potential of each species.   If one looks at a system at equilibrium, one will be unable to tell, unless one also looks at the walls and the surroundings, whether the system is isolated or not.   The relationships among the properties of a system at equilibrium do not depend on how the equilibrium is maintained, only on the fact that it is indeed an equilibrium state.   Other important equilibrium situations are those where chemical reactions are occurring; these will be studied in Sec. 30.

16   SUMMARY OF THE PRINCIPLES

This section summarizes the body of thermodynamic principles obtained in Part I of this book.   Although the derivations of these principles will not be easily remembered on just a few readings, the results can be remembered, since there really are not very many of them.   In Part II, as these principles are used over and over, they will become more intuitive, more self-evident.   Then further rereadings of Part I will take on added meaning.

A system is described by certain properties, which take their values as the results of prescribed measurements on the system. Not all properties are independently variable; there are relations between them by which some may be calculated if others are known. There exists a minimum number of independent properties for the complete description of a thermodynamic state. All other properties imaginable for the system in that state could, in principle, be computed from a knowledge of the minimum set of state variables. A major task of thermodynamic theory is to find the relationships between these properties.

The work done by the surroundings on an adiabatic system may be found from a device so made that the sole effect in the surroundings is the raising or lowering of a standard weight in the earth's gravitational field. A variety of kinds of work exist—mechanical, compressional, electrical, magnetic, gravitational, surface—but we temporarily treat only compressional or pressure-volume work. This is given by

$$W = \int dW = - \int_{\text{state 1}}^{\text{state 2}} p_{\text{ext}} \, dV \qquad \text{(integral along the particular path)}.$$

$$(16\text{-}1)$$

The work for a process taking the system from state 1 to state 2 cannot be computed by knowing just the states, since work pertains to a process, a particular path, and not just to the states.

In much the same fashion that work defines the energy function of state in mechanics, it does in thermodynamics too, as long as the system can interact with its surroundings only mechanically, that is, as long as there is no heat interaction. Energy changes are defined by

$$\Delta E = E_2 - E_1 = \int_1^2 dE = \int_1^2 dW_{\text{adiab}} = W_{\text{adiab}}. \qquad (16\text{-}2)$$

This is the content of the first law of thermodynamics.

The heat absorbed by a system in a particular process may be found from the equation

$$dE = dQ + dW, \qquad (16\text{-}3)$$

which defines heat. Integration of dQ along the particular path of interest yields

$$\Delta E = Q + W. \qquad (16\text{-}4)$$

The second law of thermodynamics, as we have stated it, says that any isolated system eventually runs down and reaches a time-independent equilibrium state. This implies the so-called zeroth law of thermodynamics, which says that things in thermal equilibrium with the same thing must prove to be in thermal equilibrium with each other. The zeroth law permits defining the temperature as that function of state common to all systems in thermal equilibrium with each other. The Kelvin temperature scale is a consistent and useful means of assigning numbers to different temperatures:

$$\theta_K = \frac{1}{R} \lim_{\rho \to 0} \frac{pV}{n} \qquad \text{(any real gas).} \qquad (16\text{-}5)$$

Here, pV/n is measured at different densities for a thermostated real gas. The limit of this function of ρ as $\rho \to 0$ is independent of the choice of gas. The gas constant R is defined to be 0.08205967 l-atm/deg-mole.

A process is reversible if each infinitesimal step of it could have been precisely reversed by an infinitesimal change in the condition of the surroundings at the time of that step. Processes are kept from being reversible if they occur too fast for the system to be nearly at equilibrium at all times or if there is friction, hysteresis, overvoltage, finite temperature differences, etc.

There exists a function of state called the entropy, defined by the equation

$$dS = \frac{dQ_{\text{rev}}}{T}, \qquad (16\text{-}6)$$

where T is a universal function of the temperature only. The Kelvin temperature is suitable for use in Eq. 16-6. For irreversible processes one can write the Clausius inequality

$$dQ \leqslant T_{\text{surr}}\, dS. \qquad (16\text{-}7)$$

The efficiency of a Carnot engine acting between two heat reservoirs is the ratio of work done by the engine in a cycle to the heat absorbed from the hot reservoir. This is $(T_{\text{hot}} - T_{\text{cold}})/T_{\text{hot}}$. For a given infinitesimal change of state in a system, maximum work will be done on its surroundings (or minimum work will be done on it) if the change is performed reversibly. Real engines, not being reversible, have lower efficiencies than Carnot engines.

The combined first and second laws for open systems are reflected in the Gibbs equation for changes in internal energy:

$$dU = T\, dS - p\, dV + \sum_k \mu_k\, dn_k, \qquad (16\text{-}8)$$

an equation which defines the chemical potential μ_k. With Eq. 16-8, the goal of relating all the system properties to a minimum set of independent properties is achieved for equilibrium systems. If one knows U as a function of S, V, and n_i, one can calculate T, p, and μ_k simply by differentiation:

$$T = \left(\frac{\partial U}{\partial S}\right)_{V,\, n_i}, \qquad p = -\left(\frac{\partial U}{\partial V}\right)_{S,\, n_i}, \qquad \mu_k = \left(\frac{\partial U}{\partial n_k}\right)_{S,\, V,\, n_j \neq k}. \qquad (16\text{-}9)$$

However, it is often convenient to work with independent variables other than S, V, and n_i. Depending on the set of independent variables chosen, thermodynamic potentials other than U afford the means of finding all the equilibrium properties from a minimum set. For the variables S, p, and n_i, one uses the enthalpy

$$H \equiv U + pV. \tag{16-10}$$

For T, V, and n_i, one uses the Helmholtz free energy

$$A \equiv U - TS. \tag{16-11}$$

For T, p, and n_i, one uses the Gibbs free energy

$$G \equiv U - TS + pV. \tag{16-12}$$

For T, p, and μ_i, one uses the potential Ω, which has the value zero for large systems:

$$0 = U - TS + pV - \sum_i \mu_i n_i \qquad \text{(large system).} \tag{16-13}$$

From Eq. 16-13 and the defining Eqs. 16-10 through 16-12, expressions for U, H, A, and G for large systems follow.

The set of Gibbs relations for the Legendre transformed potentials,

$$dH = T\,dS + V\,dp + \sum_i \mu_i\,dn_i, \tag{16-14}$$

$$dA = -S\,dT - p\,dV + \sum_i \mu_i\,dn_i, \tag{16-15}$$

$$dG = -S\,dT + V\,dp + \sum_i \mu_i\,dn_i, \tag{16-16}$$

$$0 = -S\,dT + V\,dp - \sum_i n_i\,d\mu_i \qquad \text{(large system),} \tag{16-17}$$

along, of course, with Eq. 16-8 and the defining Eqs. 16-10 through 16-12, tell us precisely how to relate the entire set of equilibrium properties to *any* suitable minimum set of properties by differentiation in the manner of Eqs. 16-9. Equation 16-17 is the Gibbs equation for the Ω potential in the limit of large systems where Ω is zero. It is called the Gibbs-Duhem equation.

It is sometimes useful to employ the Maxwell relations from Eqs. 16-15 and 16-16 to express hard-to-measure derivatives in terms of some that are relatively easy to find experimentally:

$$\left(\frac{\partial S}{\partial V}\right)_T = \left(\frac{\partial p}{\partial T}\right)_V, \qquad \left(\frac{\partial S}{\partial p}\right)_T = -\left(\frac{\partial V}{\partial T}\right)_p. \tag{16-18}$$

All parts of an equilibrium system which are in thermal contact must have the same temperature, all parts which can exchange volume without friction (omitting surface effects) must have the same pressure, and all parts which can exchange matter of type k must have the same value for μ_k. The chemical potential of substance k is its partial molar Gibbs free energy:

$$\mu_k = u_k - Ts_k + pv_k \qquad (16\text{-}19)$$

where procedures exist for measuring u_k, s_k, and v_k in mixtures.

In Part II, the whole tenor of this book changes. From concern with the grand design of a marvelous conceptual structure, we turn to using these concepts to organize, interpret, and predict a wide variety of properties and events. We now must get our hands dirty, begin to solve problems, and come up with right numerical answers. This involves worrying about details such as signs and units and the use of slide rules. It might seem like we are leaving the sublime heights to grub in the mud of details. But it is just these details and numerical calculations that give such great power to thermodynamics. The only way to develop confidence and ability to apply thermodynamics is to apply it—to practice it—over and over, to solve problems of all kinds until the insight deepens, good habits get set, and the mystery is lessened.

However, the mystery in thermodynamics, like that in a beautiful woman or a Van Gogh painting, can never be entirely dispelled. Even those people who best know and love thermodynamics get confused and make their mistakes. This book will no doubt leave its bit of confusion as well as its share of clarification. That is the way with life.

PART II

APPLICATIONS OF
THERMODYNAMICS

17 ELEMENTARY PROBLEM SOLVING

The variety of problems one might encounter in thermo-dynamics is large enough to inspire awe. It is therefore wise to develop a routine in approaching them which may be followed confidently with each problem. A consensus has arisen among people who regularly use or teach thermodynamics regarding the best routine, one which will maximize clear thinking and good results. Briefly, it is the following four steps:

The first task on reading a problem is to try to **figure out just what is happening**, what kind of process is occurring. Is it an engine operating between two thermal reservoirs, is it the isothermal expansion of a gas, is it the vaporization of a liquid at its boiling point? A small *sketch* almost always aids in visualization, even if the sketch amounts to no more than simply rewriting the data given in the problem.

Once the process is understood, one needs either to recall, or—much preferred—to **derive quickly the correct general equation** which will relate the desired quantities to what is known for the kind of process considered.

It is often tempting to wrack one's brain or hunt through a book or the literature to find the desired general equation in an effort to avoid this procedure. However, two secrets to successful study of thermodynamics are the possession of a weak memory and a strong mistrust of results derived by other people. When very little is remembered and most things are derived, mistakes are less likely to be made, or, if made, are more likely to be caught. This increases one's confidence, experience, and sophistication in approaching new problems, and increases understanding of the process being studied and the origins and limitations of the theoretical treatment being used. It is much better to refer to a book for hints in the derivation than for the final result, or to follow through the derivation in a text, understanding each step, rather than just copying the end result and trying to use it.

One should **work with general algebraic equations as long as possible** before substituting in any numbers. Thus the final result should be expressed generally, say, in terms of initial and final temperatures T_1 and T_2 or pressures p_1 and p_2, of gas-law parameters a and b, of the gas constant R, etc. Only at the last moment, and just for the purpose of getting a numerical answer, should one start substituting in the numerical values of these algebraic symbols. Do not get bogged down in arithmetic and the problems of units when you still are concerned with algebra and calculus. Only when the desired result has finally been solved for algebraically in terms of known quantities should the numerical values of the known parameters and quantities be substituted into the equation and the numerical answer obtained.

When numbers finally are substituted, **with each number should be included its units**. The units are treated just like algebraic quantities; the same unit may be canceled between numerator and denominator; terms may be added or subtracted only if they have the same units. The units of a number are every bit as important as the number; the one is meaningless without the other. The value of R is not 1.987! It may well be 1.987 cal/deg-mole, but it is also 8.315×10^7 ergs/deg-mole or 0.082 l-atm/deg-mole, or many other things. One picks the value one thinks will lead to the simplest results for the units of the answer. But once a unit accompanies a number into a calculation, it must be carried through every subsequent step until it is either canceled out by an identical unit or appears in the final numerical answer. Techniques for handling units will be acquired with experience, but their consistent careful use will cause an enormous difference in clear thinking and in getting good results.

The importance of these four steps to success in applying thermodynamics (or any science) to problem solving cannot be overstated.

Example 17.1. A byword in freshman chemistry classes is that "22.4 l of an ideal gas at 1 atm and 0°C contains 1 mole." This sentence is simply a long-winded way to remember the value of R in a particular set of units. Calculate R from it.

Solution: The problem involves simply plugging data into the ideal-gas equation, $pV = nRT$:

$$R = \frac{pV}{nT}, \tag{17-1}$$

$$R = \frac{(1 \text{ atm})(22.4 \text{ l})}{(1 \text{ mole})(273 \text{ deg})} \tag{17-2}$$

$$R = 0.082 \text{ l-atm/deg-mole.} \tag{17-3}$$

The only things to watch are the use of Kelvin degrees, *never centigrade* (Celsius), and the careful treatment of the units. There was never any question about the units of the answer. Since the correct unit was used with each numerical factor, the units of the answer fell out naturally without any further thought.

Example 17.2. A mole of water vapor at 100°C and 1 atm of pressure is compressed at constant pressure to form liquid at 100°C. The heat of vaporization of water is 540 cal/g. Compute V_1, V_2, W, Q, and ΔU (in ergs) for the system. *Given:* $R = 1.987$ cal/deg-mole $= 0.082$ l-atm/deg-mole $= 8.315$ joules/deg-mole.

Solution: The problem does not specify that the process is reversible, but Q and W can be found only if one presumes that it is. Furthermore, one is given no information about the initial volume of the water vapor, but in the context the only way to work the problem is to presume that the ideal-gas law may be used to find it. It is unfortunate to have to presume these things, but it is commonly necessary. Problems rarely are given so carefully that one need not make presumptions; making presumptions that are physically reasonable is part of the art of being a scientist:

$$V_1 = \frac{nRT}{p} = \frac{(1 \text{ mole})(0.082 \text{ l-atm})(373 \text{ deg})}{(\text{deg-mole})(1 \text{ atm})} = 30.6 \text{ l} \qquad (17\text{-}4)$$

If R had been chosen in less appropriate units, this would have been glaringly apparent in the units of the answer. For example,

$$V_1 = \frac{nRT}{p} = \frac{(1 \text{ mole})(1.987 \text{ cal})(373 \text{ deg})}{(\text{deg-mole})(1 \text{ atm})} = 742 \text{ cal/atm.} \qquad (17\text{-}5)$$

This answer is correct, it is simply an unconventional choice of units for volume!

Note: It is sometimes tempting just to look at a formula as simple as $V_1 = nRT/p$ and plug in obvious numbers, that is, $V_1 = (1)(1.987)(373)/(1) = 742$, and then to tack on some obvious units, liters, say, in the case of volume. How wrong this would be! It is an approach which is often appealing to save time in routine problems, but which makes for careless and sloppy work, and is doomed to frequent time wasting, silly mistakes, which sometimes snowball into serious conceptual errors. With the careful approach of either Eq. 17-4 or 17-5, a serious mistake is almost impossible.

We know that the liquid water must have a density not too different from 1 g/cm³, even though, again, nothing is explicitly mentioned about it in the problem. Thus

$$V_2 = (18 \text{ g})\left(\frac{1 \text{ cm}^3}{\text{g}}\right) = 18 \text{ cm}^3 \qquad (17\text{-}6)$$

In order to compute the work, a quick sketch helps in visualization:

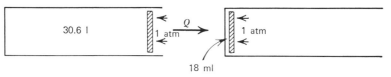

30.6 l Q 1 atm   1 atm

18 ml

The work is given by the integral of $-p \, dV$ over the constant-pressure path:

$$W = -\int_{V_1}^{V_2} p \, dV = -p \int_{V_1}^{V_2} dV = -p(V_2 - V_1), \qquad (17\text{-}7)$$

$$W = -(1 \text{ atm})(0.018 \text{ l} - 30.6 \text{ l}) = 30.6 \text{ l-atm} \qquad (17\text{-}8)$$

It is always wise to check the sign of answers in thermodynamics, where possible, by common-sense arguments. The work done on a system when the volume is squeezed down from large to small should be positive, so Eq. 17-8 has the right sign. We note that it made no difference how precisely we knew the molar volume of the liquid water; the volume of condensed water was negligible compared to the volume of the vapor; thus it never entered the answer.

The heat is calculated directly:

$$Q = \frac{-(540 \text{ cal})(18 \text{ g})}{\text{g}} = -9720 \text{ cal}. \qquad (17\text{-}9)$$

The minus sign was inserted because heat is given off by the gas as it condenses.

Now we can find $\Delta U = Q + W$. In order to add quantities, they must have the same units; in this case ergs are called for. We take advantage of the fact that multiplication of a quantity by unity leaves its value unchanged. Therefore, since 10^7 ergs = 1 joule, multiplication of anything by (10^7 ergs/1 joule) or by (1 joule/10^7 ergs) does not change its value. Also we presume that we do not remember the conversion from one energy unit to another, since this is surplus knowledge anyway. But values of R in several important units are given with the problem, and these are, in fact, well worth remembering. We can simply multiply any quantity by R/R, leaving its value unchanged, but by picking appropriate units for the R's, we can convert energy units as desired. Thus we obtain

$$\Delta U = Q + W = -9720 \text{ cal} + 30.6 \text{ l-atm} \qquad (17\text{-}10)$$

$$= \frac{(-9720 \text{ cal}) \, (\text{deg-mole})(8.315 \text{ joules})(10^7 \text{ ergs})}{(1.987 \text{ cal})(\text{deg-mole})(\text{joule})}$$

$$+ \frac{(30.6 \text{ l-atm})(\text{deg-mole})(8.315 \text{ joules})(10^7 \text{ ergs})}{(0.082 \text{ l-atm})(\text{deg-mole})(\text{joule})} \qquad (17\text{-}11)$$

$$= -40.7 \times 10^{10} \text{ ergs} + 3.11 \times 10^{10} \text{ ergs}, \qquad (17\text{-}12)$$

$$\Delta U = 37.6 \times 10^{10} \text{ ergs}. \qquad (17\text{-}13)$$

If the statement of the problem had indicated, or if its origin in a real-life situation had warranted, much greater precision in the data and, thus, in the answer could have been obtained.   The density of water vapor at 100°C and 1.000 atm of pressure could have been found in the steam tables of any chemistry handbook.   These also contain tables of precise heats of vaporization for water.   Since textbook problems need not be exercises in computational precision, most of them are satisfactorily performed to only slide-rule accuracy.

PROBLEMS

17-1. A cylinder and piston contain a gas which is maintained at *constant* pressure p.   The piston surface is a circle of radius 5.0 cm.   The piston is pushed inward reversibly by the surroundings 0.42 in.   The work required to displace the piston is 0.85 cal.   Find the pressure in atmospheres.
17-2. Do Prob. 11-13, if not already done.

18   p-V-T DATA

The theoretical framework of thermodynamics presented in Part I is general; it may be used with any macroscopic equilibrium system.   In order to apply it to a particular material, data must be introduced about the system at hand.   The two most commonly available kinds of such data are *volumetric* (relations among p, V, T, and composition) and *thermal* (measurements of heats of phase transitions and of heat capacities as functions of T).   These data represent the most accessible bridge between the general theory and the application of thermodynamics to a specific material.   Furthermore, between them they are sufficient to permit the calculation of any other properties of interest.   This section discusses typical relationships among p, V, and T for materials of fixed composition as they are found from experiment and as they are conveniently recorded and tabulated.   Heat capacities are treated in Sec. 19.   Systems in which composition is a variable are treated in Secs. 26 to 30.

THE EQUATION OF STATE

A *homogeneous* system is one which has uniform intensive properties throughout. A *phase* is a homogeneous system (or subsystem). A *heterogeneous* system has two or more phases, each of which is usually open to exchange of matter with the other(s). A single phase may exist in several disjoint pieces; thus a million fog droplets suspended in air represent a two-phase system, the water droplets making up one and the air making up the other.

We postpone the consideration of chemical reactions until Sec. 30. If no chemical reactions can occur, then each distinct chemical species present is called a *component* of the system, the number of components c being equal to the number of such species. As we saw in the discussion of Sec. 13, any set of $c + 2$ independent variables will characterize an equilibrium system. Such a set is V, T, and n_i; thus the pressure at equilibrium must be a function of V, T, n_i:

$$p = p(T, V, n_i). \qquad (18\text{-}1)$$

This relationship is called the *equation of state* for the material under consideration, or sometimes the *thermal* equation of state, to distinguish it from the so-called *caloric* equation of state: $U = U(T, V, n_i)$. It sometimes is written in the more symmetric form (see Eqs. A-1 to A-3 in Appendix A),

$$f(p, V, T, n_i) = 0. \qquad (18\text{-}2)$$

Since temperature, pressure, volume, and mass are relatively easy quantities to measure with great precision, p-V-T data are among the most precise information available about thermodynamic systems. There are a great many ways in which these data are recorded and expressed. Tables of raw data are often the only way they can be reported accurately. However, tables are difficult to use in numerical calculations and hard to develop physical feeling for. Thus one commonly resorts to either graphical methods to record the data or to mathematical approximations showing the relationships among p, V, T, and n_i.

EXPERIMENTAL p-V-T DATA

Graphical representations of p-V-T data have the advantage of conveying physical insight about the equations of state. They facilitate interpolations among the experimental data. If one wants simply to find one of the three—p, V, or T—when the other two are given, graphs are excellent. However, if one wants to use the relationship among p, V, and T to find the values of integrals or derivatives of these functions, a mathematical approximation that fits the data and does the interpolation analytically is much more useful. Even tables of raw data are superior to graphs for these purposes, since the integration and differentiation can be done numerically, procedures which have been greatly facilitated by high-speed computers.

If pressure is plotted against molar volume and temperature for 1 mole of a typical pure substance, the result commonly resembles Fig. 18-1. The

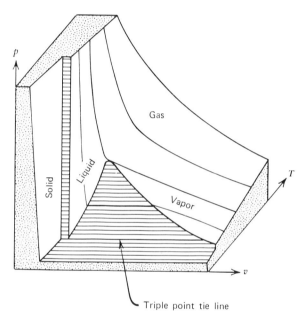

Fig. 18-1. Plot of p versus v and T for 1 mole of a typical substance (scale distorted for clarity).

white surface is the p-V-T surface; given any two of the three coordinates, the other is found from its value on the surface. A few lines of constant temperature *(isotherms)* have been drawn in the fluid region to show the shape of the surface there. The figure may be easier to visualize if one compares it with Fig. 18-2, which is a picture of Fig. 18-1 taken parallel to the T-axis, that is, projected onto the p-v plane.

First, let us focus on the solid surface. The **solid** is loosely defined as that state of matter which has for a given amount of material both a definite volume and a definite shape. The solid surface does not differ too much from a plane, sloping almost vertically downward as one goes to higher molar volumes and sloping steeply upward as one goes to higher temperatures.

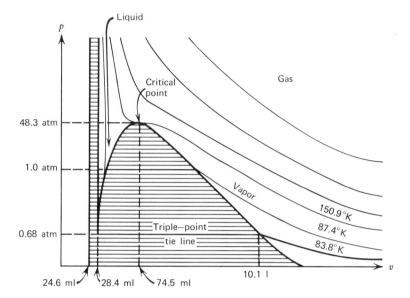

Fig. 18-2. Two-dimensional projection of Figure 18-1 onto the *p-v* plane. The values given are for argon, showing the extent to which the scale of the diagram has been distorted for the sake of clarity.

This is familiar from our experience, since most solids are highly incompressible, which means that an extraordinarily large pressure is required to decrease their volume very much.

We know this incompressibility is caused on the molecular level by the fact that in the solid the molecules (or atoms) making up the material are packed very closely together. Each molecule interacts with nearby ones with a potential energy which varies with the distance separating the two, roughly as shown in Fig. 18-3. Since the force the particles exert on each other is the negative slope of the interaction potential, as the particles get so close together that their electronic orbitals begin to overlap, they begin to repel each other strongly. Thus for all practical purposes, one can think of each particle as having a hard-core region, which excludes all other particles from that space, surrounded by an attractive potential well. Since in the solid the particles are packed so closely together, compressing them further means the pressure must be drastically increased in order to squeeze them still closer against the steeply repulsive potentials. The solidity of solids is due to the attractive wells. Individual particles are bound to a number of their neighbors in the lattice by the attractive forces, and at the temperatures where solids exist there is insufficient thermal energy to break a group of particles loose from their constraining neighbors so they can slip past each other. Thus solids hold their shapes over ordinary experimental time periods under

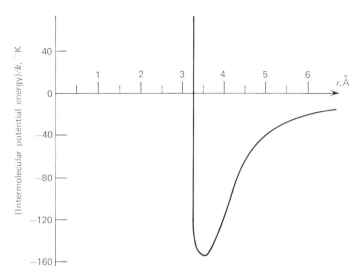

Fig. 18-3. Potential energy of a typical pair of molecules as function of distance of separation r (energy is divided by Boltzmann's constant $k = R/N_0$; molecule chosen is argon).

normally applied stressing forces. (Solids flow very much like liquids if they are subjected to large enough stresses and if they are observed on a sufficiently long time scale.)

Next we note the cross-hatched surface separating the solid and liquid regions. If p, v, and T should happen to lie on that surface, the system exists at equilibrium in two different coexisting phases, solid and liquid. The molar volumes of the two phases are given by the end points of the horizontal line (called a **tie line**) through the point representing T, p, and the overall or average molar volume. Thus the coexistence surface is filled with horizontal tie lines coupling the molar volumes of the solid and liquid phases which would be in equilibrium at the particular temperature and pressure of the line. Typical materials show the property, reflected in Fig. 18-1, of expanding slightly on melting. Water and a few metals (like antimony) show a volume *decrease* on melting, thus are abnormal in this respect.

The other cross-hatched surfaces of Fig. 18-1 represent the coexistence regions between solid and vapor and between liquid and vapor phases. Again, these surfaces are filled with tie lines, so if the p-v-T point representing the overall system falls on one of them, the system must exist not in a single equilibrium phase, but must split into two phases at that same T and p, but with the two molar volumes represented by the end points of the tie line. Clearly, the molar volume of the vapor is much, much greater than that of the condensed phase, either liquid or solid. The values given in Fig. 18-2 for argon show the extent to which the scale of the figures has been distorted for the sake of clarity. We note that there is one particular pair of T, p values for which all three phases can coexist, with any relative amounts of the three phases present, all at equilibrium. This is called the **triple point** for the substance under consideration. The triple point of water is a reproducible reference situation whose temperature has been chosen to be 273.1600°K, which defines the size of the degree in the Kelvin temperature scale, as discussed in Sec. 14.

We note that there is a maximum pressure and temperature above which there is no coexistence region for liquids and vapors. This maximum, representing the very top of the humped coexistence region, is called the **critical point**, defining the **critical pressure** p_c, the **critical temperature** T_c, and the **critical** (molar) **volume** v_c. The state characterized by the critical point is called the **critical state**, and the isotherm through the critical point is the **critical isotherm**. Perhaps the most striking feature of Fig. 18-1 is that above the critical isotherm there are not separate surfaces for liquids and vapors, but there is just one single fluid surface (fluid meaning something that flows). This bears some discussion.

The **liquid state** is usually defined to be that which has a fixed volume for a given amount of material, but not a fixed shape. It will alter its shape to fit the bottom of any vessel in which it finds itself. The **gas**, on the other hand, is that state of matter for which a given amount of material has neither a definite volume nor shape, but which expands to fill whatever container it is put in. The word **vapor** is often used for a gas at any temperature below the critical temperature, though the distinction between gas and vapor is often ignored by various writers. Obviously these definitions are arbitrary. For example, they say nothing about the time scale on which these changes are to take place. Thus a piece of glass, which has many of the features of a liquid, flows at room temperature on a time scale long compared to human observation times. Its flow, however, may be seen in the fact that large glass windows which have stood upright in buildings for long times become thicker at the bottom than at the top.

In the liquid region of Fig. 18-1, since the molar volume is small and the density is high, the repulsive part of the intermolecular potentials plays a great role. Therefore liquids are hard to compress, and the liquid surface slopes almost as steeply downward with increasing v as the solid surface does. Of course in liquids there is enough thermal energy to break down the crystal lattice and permit the molecules to slide over each other. However, the entire liquid mass does hold together, because there is so little thermal energy that it is only a rare, extraenergetic molecule which can break out of the potential wells of its neighbors and join the vapor.

In the gaseous region of Fig. 18-1, the behavior, as one goes to higher and higher v (that is, $\rho = 1/v \to 0$) approaches more nearly the ideal behavior, $pV = nRT$. The behavior of the fluid clearly departs from ideality as the density increases, where the effects of the intermolecular potentials become pronounced.

Since the two phases, liquid and gas, can coexist only in the cross-hatched region, it is possible to pass from a state clearly in the vapor region to one clearly in the liquid region without ever noticing a phase change. To get the vapor, suppose we start with 1 mole of a liquid in the bottom of a sealed evacuated vessel at its boiling point and boil it until it has completely changed to vapor. We now change it back to liquid without a discontinuous process of condensation. We fix the volume and heat the vapor to some temperature above T_c. Next we put the vessel in a thermostat at this high temperature so that T remains constant, and compress the gas to the molar volume of the liquid we started with. During this compression, there is no change of phase; a homogeneous, single-phase system is always present in the vessel. If we now fix the volume and simply cool the vessel to the original temperature, we again see no phase change in the system, yet we can open the vessel and pour out a liquid, the same liquid we started with. It passed from a state that was clearly vapor to a state that was clearly liquid by going around the critical point and thus never undergoing a discontinuous change of phase.

In light of this it seems impossible really to tell which fluid phase one has as long as it fills the containing vessel. In fact, the distinction between liquid and gas as defined previously is arbitrary for all situations except those in which both liquid and gas coexist. In those cases, the liquid property of having fixed volume is clearly demonstrated because the gas is filling up the rest of the container. In order to preserve the utility of the two words, however, many people define **liquids** to be those fluids whose temperature is below T_c and whose molar volume is less than v_c (that is, $T < T_c$ and $\rho > \rho_c$); all other fluids they call **gases**.

LAW OF CORRESPONDING STATES

A convenient method of expressing p-V-T data is through tables or graphs of the so-called **compressibility factor** z defined by

$$z \equiv \frac{pV}{nRT}, \tag{18-3}$$

given as a function of either V and T or of p and T. If the gas were ideal, z would be unity, so departures from $z = 1$ represent departures from ideality in the gas behavior. In the limit of vanishing density, z for all gases approaches unity.

A technique for handling p-V-T data that has been especially useful has been the recognition that, to quite a good approximation, *the shape of the fluid surface in Fig. 18-1 is the same for a great many fluids.* Of course, the *scale* of the p, V, and T axes differs for each fluid, since the critical points of various substances range over widely differing conditions. For example, for hydrogen, $T_c = 33.3°K$ and $p_c = 12.8$ atm, while for CCl_4, $T_c = 556.3°K$ and $p_c = 45.0$ atm. However, if p, v, and T are measured in so-called **reduced units**, defined by

$$p_r = \frac{p}{p_c}, \qquad T_r = \frac{T}{T_c}, \qquad v_r = \frac{v}{v_c}, \qquad (18\text{-}4)$$

then the fluid surfaces in Fig. 18-1 for a wide variety of fluids could be super-imposed with surprisingly little error. This is called the **law of corresponding states**, and it may be written mathematically in the form

$$p_r = p_r(T_r, v_r) \qquad \text{(same function for many fluids)}. \qquad (18\text{-}5)$$

There are only three parameters in Eq. 18-5 which might vary from one fluid to another (p_c, T_c, and v_c), but one of these is fixed by knowing that, as $v_r \rightarrow \infty$, Eq. 18-5 must approach $p = RT/v$ (which fixes the gas constant R). Thus all fluids which "correspond" obey the same equation of state, an equation containing only two parameters which can be adjusted to fit the particular fluid.

The molecular basis of this law lies in the fact that the intermolecular potentials of a great many kinds of molecules have the same general shapes. The potential function, Fig. 18-3, for a host of fluids may be represented accurately by an analytic formula containing only two parameters whose values may be adjusted to fit the particular substance at hand. These fluids would be expected to obey the law of corresponding states.* If three parameters were needed to represent the intermolecular potentials (which is always required if the molecules are not spherical), then three would be needed to express the equation of state. For these fluids, instead of Eq. 18-5, one has

$$p_r = p_r(T_r, v_r, \omega) \qquad \text{(same function for a great many fluids)}, \qquad (18\text{-}6)$$

where ω is one more parameter which is fit for the substance at hand. Several choices of a third parameter have been made, including the value z_c of the compressibility factor at the critical point. There is now limited agreement

---

* See, for example, F. C. Andrews, *Equilibrium Statistical Mechanics*, John Wiley & Sons, New York, 1963, p. 165.

that the best choice for ω is a function of the vapor pressure at $T = 0.7T_c$ called the *acentric factor*.*

A convenient way to express p-V-T data for fluids that approximate the law of corresponding states is to plot lines of constant T_r on a z versus p_r graph, as shown in Fig. 18-4. It is easy to understand the features of Fig. 18-4 qualitatively in terms of the intermolecular potentials of the constituent molecules. At high temperatures, say, greater than $2\frac{1}{2}$ times T_c, the attractive wells of the molecules' potentials represent very small perturbations

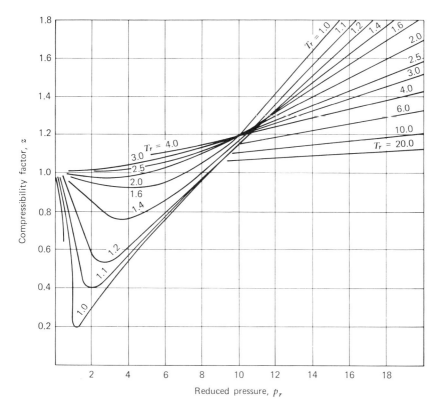

Fig. 18-4. Compressibility factor for various constant reduced temperatures plotted against reduced pressure. Figure is useful for a great many fluids that obey the law of corresponding states. [L. C. Nelson and E. F. Obert, *Trans. A.S.M.E.*, **76**, 1057 (1954).]

* For the use of z_c, see O. A. Hougen, K. M. Watson, and R. A. Ragatz, *Chemical Process Principles: Part II, Thermodynamics*, 2d ed., John Wiley & Sons, New York, 1959, chapter 14. For the use of the acentric factor, see G. N. Lewis, M. Randall, K. S. Pitzer, and L. Brewer, *Thermodynamics*, McGraw-Hill Book Co., 2d ed., New York, 1961, appendix I.

to the typical thermal energy which the molecules have. Therefore, at so high a temperature only the hard cores of the molecules play significant roles. Thus the departures from ideality are increases in the pressure due to the fact that the volume the molecules have in which to rattle around is smaller than V by the amount occupied by the hard cores. However, as the temperature is lowered, the significance of the attractive well increases in relation to the average thermal energy that molecules possess. When the temperature is low enough, the thermal energy of the molecules can be overcome by the attractive intermolecular forces, and a significant decrease in pressure over the ideal behavior results. One may think of this as caused by some pairs of molecules being more or less stuck together, so that the actual number of freely flying particles in the gas is less than the number of molecules. Of course, after the pressure gets high enough so that the molecules begin to get squeezed together significantly, the excluded volumes of the hard cores predominate over the attractive wells and the pressure increases rapidly.

The difference in scales for various gases is noteworthy. For hydrogen, the gas would approximate the isotherm $T_r = 1.0$ at $33.3°K$, while carbon tetrachloride would approximate this isotherm at $283.1°C$. At room temperature, hydrogen already has a reduced temperature of about 9, while CCl_4 has T_r just over one-half. A reduced pressure for 10.0 for hydrogen represents an actual pressure of 128 atm, while $p_r = 10.0$ for carbon tetrachloride means an actual pressure of 450 atm. Given the table of critical constants for gases that is available in any chemical handbook and a graph of the type of Fig. 18-4, one would be able to estimate either p, V, or T to a reasonable accuracy, given the other two. Where greater accuracy is demanded, the use of a third adjustable parameter permits minor corrections in the answers obtained, and results good to 2% can consistently be obtained (except for highly polar or strongly associated hydrogen bonded substances).*

ANALYTIC EQUATIONS OF STATE

Since one often has to perform mathematical operations on certain of the functional relationships among p, V, and T, it is useful to represent the equation of state analytically. Well over 100 of such analytical equations have been proposed in the scientific literature. Unfortunately, it has proved

---

* References to these methods may be found in W. R. Gambill, *Chem. Eng.*, **66** (21), 195 (1959). Excellent graphs are presented by L. C. Nelson and E. F. Obert, *Trans. A.S.M.E.*, **76**, 1057 (1954). The three-parameter methods are compared with experiment by E. D. Hooper and J. Joffe, *Chem. Eng. Data Series*, **5**, 155 (1960).

very difficult to construct one which is useful at pressures much above p_c or at liquid densities.*

The easiest analytic equation of state is the ***ideal-gas law***

$$pV = nRT \tag{18-7}$$

which has the same form for all gases. This equation may be derived from molecular theory for gases consisting of either classical mechanical or quantum mechanical molecules, as long as the effects of the intermolecular forces on the properties of the gas are completely negligible. This occurs only in the limit of vanishing density, but, in fact, the ideal-gas law is useful at all pressures up to p_c if the temperature is at least two or three times T_c. Otherwise it is highly inaccurate. In Fig. 18-2, one can observe how the isotherms approach the hyperbolas of Boyle's law as the temperature increases, and particularly as the pressure is decreased.

From the law of corresponding states one would conclude that at least two adjustable parameters are needed for an analytic equation to be at all faithful to the data. Perhaps the best known two-parameter equation is that proposed in 1873 by J. D. ***van der Waals***:

$$\left(p + \frac{n^2 a}{V^2}\right)(V - nb) = nRT. \tag{18-8}$$

The van der Waals equation has a none too convincing justification based on the molecular picture and is easy to manipulate mathematically. The parameters a and b are assigned to each different chemical compound by one of several methods described below. Once these values have been chosen, the equation relates p, V, and T, supposedly over the entire fluid surface in Fig. 18-1. The fit between experiment and equation is good in the gaseous region for $T_r < 1$, it is also good up to $p_r = 1$ for $T_r > 1$, and by the time T_r gets up to about 2, the equation is good up to reduced pressure of about 8. For higher pressures and in the liquid region, the van der Waals equation is unsatisfactory.

The three most commonly encountered two-parameter modifications of the van der Waals equation are the following: The equation of ***Berthelot***,

$$\left(p + \frac{n^2 a}{TV^2}\right)(V - nb) = nRT \tag{18-9}$$

is similar to van der Waals' equation, with a different temperature dependence for the pressure correction. It represents no improvement, however, and, if

---

* For critical reviews of various equations of state, see J. Joffe, *Chem. Eng. Progr.*, **45**, 160 (1949), and K. K. Shah and G. Thodos, *Ind. Eng. Chem.*, **57** (3), 30 (1965).

anything, a decrease in accuracy over van der Waals' equation. The equation of **Dietirici**,

$$p(V - nb) = nRTe^{-na/RTV} \qquad (18\text{-}10)$$

seems to be worse than van der Waals' in all regions and no good at all for $p_r > 1$. However, some authors claim that, if used as an interpolation formula in restricted regions of temperature and volume, it seems somewhat more accurate than van der Waals'. The equation proposed by **Redlich** and **Kwong**[*]

$$\left[p + \frac{n^2 a}{T^{1/2} V (V + nb)} \right] (V - nb) = nRT \qquad (18\text{-}11)$$

has proved to be surprisingly accurate. For argon it is excellent up to $p_r = 50$ for all reduced temperatures between 0.9 and 2.0. There is a small deviation from experiment in its predicted critical isotherm between $p_r = 1$ and $p_r = 4$, but striking agreement with experiment over the rest of the pressure range from $p_r = 0$ to $p_r = 50$. For *n*-butane, a less spherical molecule than argon, the agreement with experiment is still excellent, though not quite so remarkable. The Redlich-Kwong equation is a simple, accurate, two-parameter equation, useful over wide ranges of temperature and pressure.

For fluids of constant composition, it is true that, if enough volumetric data are available, the more adjustable parameters in an equation of state, the better it is, *provided* the equation is used only as an interpolation formula in the range of the data. A number of equations have been proposed containing from three up to eight constants, but in general they are not as good as Eq. 18-11 and are far more complicated. An extension of the Redlich-Kwong equation to include the acentric factor as a third parameter[†] has resulted in a complicated, but extremely accurate, three-parameter equation of state.

The parameters *a* and *b* that appear in Eqs. 18-8 to 18-11 are not necessarily related to each other; in fact, some of them do not even have the same dimensions. Once one has chosen the equation one will use, one fits " best " values of *a* and *b* for that equation and the particular fluid one is treating. If the equation is being used simply as an interpolation formula for data in a relatively small region of the *p*-*V*-*T* surface, then the parameters are chosen to make the equation best fit the available data. It then is unimportant how well the equation would represent the data outside the region of interest. If, on the other hand, one wants the equation to be as good as possible over the

---

[*] O. Redlich and J. N. S. Kwong, *Chem. Rev.*, **44**, 233 (1949).
[†] O. Redlich, F. J. Ackerman, R. D. Gunn. M. Jacobson, and S. Lau, *I. & E.C. Fundamentals* **4**, 369 (1965).

entire fluid p-V-T surface, one fits the critical point as calculated from the equation with the experimental critical point data. Critical temperature, pressure, and density are tabulated for a variety of compounds in most chemical handbooks. From the density and the molecular weight one can easily compute the critical molar volume v_c. The method of fitting the parameters is to note from Fig. 18-2 that at the critical point the critical isotherm has an inflection point. Thus

$$\left(\frac{\partial p}{\partial v}\right)_T = 0, \qquad \left(\frac{\partial^2 p}{\partial v^2}\right)_T = 0 \qquad \text{(at critical point).} \qquad (18\text{-}12)$$

The equation of state may be differentiated and the appropriate derivatives set to zero; thereby equations may be found which relate the parameters a and b to the tabulated critical constants.

Example 18-1. What are the van der Waals parameters a and b in terms of the properties of the fluid at its critical point, p_c and T_c?

We rewrite van der Waals' equation for 1 mole and differentiate it twice:

$$p = \frac{RT}{v - b} - \frac{a}{v^2}, \qquad (18\text{-}13)$$

$$\left(\frac{\partial p}{\partial v}\right)_T = -\frac{RT}{(v - b)^2} + \frac{2a}{v^3} = 0 \qquad \text{(at critical point),} \qquad (18\text{-}14)$$

$$\left(\frac{\partial^2 p}{\partial v^2}\right)_T = \frac{2RT}{(v - b)^3} - \frac{6a}{v^4} = 0 \qquad \text{(at critical point).} \qquad (18\text{-}15)$$

Now, with Eqs. 18-14 and 18-15 valid at the critical point and Eq. 18-13 valid everywhere, algebraic manipulation permits expressing the van der Waals parameters in terms of the critical properties, or vice versa. Perhaps the easiest method is to eliminate a between Eqs. 18-13 and 18-14 to find b and then to solve for a by substitution. The result is

$$a = \frac{27RbT_c}{8} = 27b^2 p_c = \frac{27R^2 T_c^2}{64p_c} \qquad (18\text{-}16)$$

$$b = \frac{RT_c}{8p_c} = \frac{v_c}{3} \qquad (18\text{-}17)$$

From these expressions and a table of critical constants one can immediately determine the van der Waals parameters for the material of interest, and then, using van der Waals' equation, make reasonable estimates of the p-V-T relationships over much of the p-V-T surface.

The van der Waals b is just one-third the critical volume; and molecular theories indicate that it is a rough measure of the actual volume excluded by the hard cores of Avogadro's number of molecules. Whenever $V \gg \frac{1}{3}V_c$, the term involving b in van der Waals' equation can be neglected. The Redlich-Kwong b parameter is related to molecular size in a similar fashion. The van der Waals' a (as with the Redlich-Kwong a) is a measure of the strength of the attractive intermolecular forces. Whenever $TV \gg T_c V_c$, the correction involving a may be neglected.

THE VIRIAL EQUATION

One approach to the equation of state of gases that is interesting ᴏoth theoretically and practically is an expansion of the compressibility factor in powers of the molar density, called the *virial equation of state*:

$$\frac{pV}{nRT} = z = 1 + B(T)\frac{n}{V} + C(T)\frac{n^2}{V^2} + D(T)\frac{n^3}{V^3} + \cdots. \qquad (18\text{-}18)$$

The $B(T)$ is called the *second virial coefficient*, $C(T)$ the *third virial coefficient*, etc. (The "first virial coefficient" turned out to be the gas constant R, so the expression is never used now.) The virial coefficients are *functions of T only* for pure substances, different functions, of course, for different substances. They are found experimentally from low-pressure *p-V-T* data by rewriting Eq. 18-18 as follows, where $\rho \equiv n/V$ is the molar density:

$$\frac{1}{\rho}\left(\frac{p}{\rho RT} - 1\right) = B + C\rho + \cdots. \qquad (18\text{-}19)$$

If the left-hand side of Eq. 18-19 is plotted against ρ for fixed T, the intercept at $\rho = 0$ is the value of B, and the slope of the curve at $\rho = 0$ is the value of C at that temperature.

The form of the function $B(T)$ is similar to the plot shown in Fig. 18-5 for argon. For most gases, $B(T)$ may be approximated by

$$B(T) = b - \frac{a}{RT} \qquad (18\text{-}20)$$

where a and b are the same as the van der Waals parameters. The negative region of $B(T)$ is caused by the attractive part of the intermolecular potential well, and the positive region by the excluded volumes of the hard cores. The maximum in the actual $B(T)$ (at around 20 in Fig. 18-5) curve stems from the fact that the hard cores of the molecules are not absolutely impenetrable. Measurements of third virial coefficients are difficult because of the great precision required of the volumetric measurements at low densities.

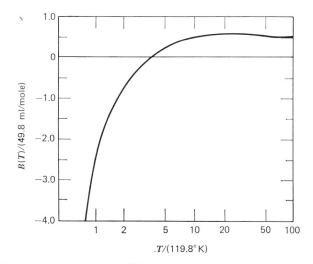

Fig. 18-5. Plot of second virial coefficient against temperature for argon.   Reference: J. O. Hirschfelder, C. F. Curtiss, and R. B. Bird, *Molecular Theory of Gases and Liquids*, John Wiley & Sons, New York, 1954, p. 164).

In statistical mechanics it is possible to derive formulas expressing the various virial coefficients in terms of the intermolecular potential energy curve, Fig. 18-3.   Unfortunately, virial coefficients are hard to compute from these formulas because the potentials are not yet well enough known.   However, the theoretical relationship between virial coefficients and intermolecular potential is a unique advantage of this equation of state over more empirical equations.

The virial equation is easy to handle mathematically and can be reasonably accurate.   For $p_r < \frac{1}{2}$, Eq. 18-18 can usually be truncated after the second virial term (that is, C, D, etc., can be neglected).   If it is truncated after the third virial term, it is satisfactory for pressures up to p_c if the temperature is greater than T_c.   For higher pressures, higher coefficients would be needed, and at present they are not available.   In the liquid region and in the region of the critical point, the whole series fails to converge.

There is an alternative virial expansion in powers of the pressure rather than the molar density:

$$pV = nRT + nB'(T)p + nC'(T)p^2 + nD'(T)p^3 + \cdots. \qquad (18\text{-}21)$$

If Eqs. 18-18 or 18-21 are truncated after the second virial term, Eq. 18-21 is often more accurate at higher densities.   If truncated after the third virial term, Eq. 18-18 is usually more accurate.

COMPRESSIBILITIES AND EXPANSION COEFFICIENTS

One form in which p-V-T data are sometimes expressed, especially for condensed phases (liquids and solids), is as the (isobaric) **coefficient of thermal expansion** α:

$$\alpha = \alpha(p, T, n_i) \equiv \frac{1}{V}\left(\frac{\partial V}{\partial T}\right)_{p, n_i}, \tag{18-22}$$

which is the fractional change in volume, dV/V, per unit temperature change (holding the pressure constant). The data are also expressed as the negative of the fractional change in volume per unit pressure change, called the (isothermal) **coefficient of compressibility** β:

$$\beta = \beta(p, T, n_i) \equiv -\frac{1}{V}\left(\frac{\partial V}{\partial p}\right)_{T, n_i} \tag{18-23}$$

The minus sign is introduced in order to make β a positive quantity, since V gets smaller when p gets larger, so $(\partial V/\partial p)_T$ must be negative.

In terms of these properties and the partial molar volumes, Eq. 12-8, changes in p, V, T and n_i may be related from Eq. 12-5:

$$dV = \left(\frac{\partial V}{\partial T}\right)_{p, n_i} dT + \left(\frac{\partial V}{\partial p}\right)_{T, n_i} dp + \sum_i \left(\frac{\partial V}{\partial n_i}\right)_{T, p, n_{j \neq i}} dn_i, \tag{18-24}$$

$$\frac{dV}{V} = \alpha\, dT - \beta\, dp + \frac{1}{V}\sum_k v_i\, dn_i. \tag{18-25}$$

For some solids and liquids, α and β may be treated as nearly constant if the ranges of temperature and pressure are not too great. In that case, integration between states 1 and 2 yields

$$\ln \frac{V_2}{V_1} \approx \alpha\, \Delta T - \beta\, \Delta p \qquad \text{(fixed } n_i\text{, small } \Delta T \text{ and } \Delta p\text{, condensed phases)}. \tag{18-26}$$

If the change in V is small compared to V itself, then integration of Eq. 18-25 yields simply

$$\frac{\Delta V}{V} \approx \alpha\, \Delta T - \beta\, \Delta p \qquad \text{(fixed } n_i\text{, all changes small, condensed phases)}. \tag{18-27}$$

For liquids, the best equation of state is that of **Tait**:

$$-\left(\frac{\partial V}{\partial p}\right)_T = \frac{K}{L(T) + p}, \tag{18-28}$$

in which K is constant and L is a function of temperature.   Integration of Eq. 18-28 at constant T yields

$$-\int_{V_1}^{V_2} dV = \int_{p_1}^{p_2} \frac{K\,dp}{L+p} \qquad (T \text{ constant}), \qquad (18\text{-}29)$$

$$V_1 - V_2 = K \ln \frac{p_2 + L}{p_1 + L} \qquad (T \text{ constant}). \qquad (18\text{-}30)$$

This equation gives remarkable agreement with experiment for a variety of liquids over a wide range of temperatures and pressures up to 1000 atm. For higher pressures, up to 25,000 atm, a modification of the Tait equation gives excellent results.*

CALCULATION OF p-V WORK

The work done on a system when its volume changes under an external pressure is computed from the integration of $dW = -p_{ext}\,dV$ over the particular path taken by the system.   As discussed in Sec. 5, p_{ext} may be replaced by the system pressure p whenever the process is sufficiently slow and either there is no friction or else the friction occurs only in the surroundings. In this case,

$$W = -\int_{V_1 \text{ path}}^{V_2} p(V)\,dV \qquad (\text{slow process}). \qquad (18\text{-}31)$$

The task is to determine what the function $p(V)$ is over the path taken by the system.

Example 18-2. How much work is required to compress 5.00 g of oxygen gas reversibly from 10 to 2 l at the constant temperature of 20°C?   Treat the oxygen as an ideal gas.

Solution: A quick sketch helps in the visualization of the problem.

10 l 5 g O_2 20°C	$\xrightarrow[T\text{ constant}]{\text{Reversible}}$	2 l 5 g O_2 20°C

Derivation of the equation to be used is from Eq. 18-31.   The desired

---

* J. O. Hirschfelder, C. F. Curtiss, and R. B. Bird, *Molecular Theory of Gases and Liquids*, John Wiley & Sons, New York, 1954, p. 261, and references therein.

independent variables in the function $p(V)$ are T and V, since these are the quantities given in the problem:

$$W = - \int_{V_1 \; T \; \text{const}}^{V_2} p(V) \, dV, \tag{18-32}$$

$$= - \int_{V_1 \; T \; \text{const}}^{V_2} \frac{nRT}{V} \, dV, \tag{18-33}$$

$$= -nRT \int_{V_1}^{V_2} \frac{dV}{V}, \tag{18-34}$$

$$W = -nRT \ln \frac{V_2}{V_1} = nRT \ln \frac{V_1}{V_2} \quad \text{(reversible, } T \text{ constant, ideal gas).} \tag{18-35}$$

This result can be derived quickly when needed without risking forgetting a term or a sign.   In Eq. 18-33 the ideal-gas law was used for $p = p(T, V)$.   In Eq. 18-34, the constants and T (also constant over the path) were taken outside the integral.   We note that the result is positive whenever $V_1 > V_2$; that is, it takes work to squeeze a system into a smaller volume.   Whenever $V_1 < V_2$, the result is negative, meaning the system does work on the surroundings.

Finally, we can substitute numbers into the result:

$$W = \frac{(5 \text{ g})(\text{mole})(1.987 \text{ cal})(293 \text{ deg})(2.3)}{(32 \text{ g})(\text{deg-mole})} \log \frac{10}{2}, \tag{18-36}$$

$$W = 147 \text{ cal.} \tag{18-37}$$

We have noted that $\ln z = 2.303 \log z$.   We arbitrarily put the answer in the units of calories, since the question did not demand any particular units.

Example 18-3.   Liquid of constant compressibility β is compressed reversibly from V_1 to V_2 at constant temperature.   The initial pressure is p_0. Calculate W.

Solution:
$$W = - \int_{V_1}^{V_2} p(V) \, dV. \tag{18-38}$$

The relationship between p and V along this path is found from integrating Eq. 18-25 at fixed T and n_i:

$$\int_{V_1}^{V} \frac{dV}{V} = -\beta \int_{p_0}^{p} dp, \tag{18-39}$$

$$\ln \frac{V}{V_1} = -\beta(p - p_0). \tag{18-40}$$

This can be solved for $p(V)$ and the result used in Eq. 18–38:

$$W = -\int_{V_1}^{V_2} \left(p_0 - \frac{1}{\beta} \ln \frac{V}{V_1}\right) dV, \tag{18-41}$$

$$= -p_0 \int_{V_1}^{V_2} dV + \frac{V_1}{\beta} \int_1^{V_2/V_1} \ln x \, dx, \tag{18-42}$$

$$= -p_0 \left. V \right]_{V_1}^{V_2} + \frac{V_1}{\beta} \left[x \ln x - x \right]_1^{V_2/V_1}, \tag{18-43}$$

$$= -p_0(V_2 - V_1) + \frac{V_1}{\beta} \left[\frac{V_2}{V_1} \ln \frac{V_2}{V_1} - \frac{V_2}{V_1} + 1 \right], \tag{18-44}$$

$$W = \left(p_0 + \frac{1}{\beta}\right)(V_1 - V_2) - \frac{V_2}{\beta} \ln \frac{V_1}{V_2}. \tag{18-45}$$

In Eq. 18-42, the new integration variable $x = V/V_1$ was defined; in Eq. 18-43, the integrals were performed; in Eq. 18-44, the limits of integration were inserted (recognizing that the logarithm of unity is zero); and in Eq. 18-45, the result was simplified.

PROBLEMS

18-1. Ten liters of air at 1 atm pressure and 0°C is separated into its pure constituent gases (air is 20 mole percent oxygen and 80 mole percent nitrogen). The oxygen and nitrogen are each stored in 5-l bulbs at 30°C. What are the pressures in the two bulbs?

18-2. Use Fig. 18-4 to determine the molar volume of nitrogen at 400 atm and 0°C. The critical pressure of N_2 is 33.5 atm and the critical temperature is 126°K.

18-3. Check each step of the derivation of Eqs. 18-16 and 18-17 from Eq. 18-13.

18-4. Equation 18-35 is fine if the initial and final states are given in terms of volumes. What would the equation be if these states were given in terms of pressures?

18-5. One hundred grams of nitrogen gas at 25°C and 760 mm is expanded reversibly and isothermally to a pressure of 100 mm. Calculate the work, assuming that the gas is adequately described by the ideal-gas law.

18-6. Suppose the point v representing the system volume is on a tie line connecting the volumes v_1 and v_2 of two different phases. Prove the so-called "lever rule:" $x_1/x_2 = (v_2 - v)/(v - v_1)$, where x_i is the fraction of material in the system having molar volume v_i.

18-7. Compute the density in moles/l and in g/cm^3 of water vapor at its boiling point under 1 atm of pressure. Treat it as an ideal gas.

18-8. If 0.955 g of O_2 gas occupies a vessel, the pressure is the same as that exerted by 2.36 g of an unknown gas in the same vessel at the same temperature. What is the molecular weight of the unknown gas if both it and O_2 may be treated as ideal?

18-9. Tabulated critical constants for chlorine are $T_c = 144.0°C$, $p_c = 76.1$ atm. Calculate the van der Waals parameters for chlorine.

18-10. Using the calculated values of a and b obtained in Prob. 18-9, estimate the pressure in a 25-l vessel containing 10 kg of chlorine gas (Cl_2) at 200°C.

18-11. The critical density of chlorine is listed as 0.573 g/cm³. Check this value against the van der Waals value, Eq. 18-17.

18-12. Prove that each of the empirical equations of state, Eqs. 18-8 through 18-11, reduces to $pV = nRT$ in the limit of vanishing density.

18-13. Prove that $B(T) = B'(T)$ for the second virial coefficients of Eqs. 18-18 and 18-21.

18-14. Consider the following three paths between states 1 and 2 to be approximately reversible. What is W in calories for each of the three paths? Hydrogen may be described by the ideal-gas law. Let state 1 be 1 mole of H_2 gas at 100°C and 1 atm. Let state 2 be 1 mole of H_2 gas at $-173°C$ and 0.01 atm. Let path 1 be the following: expand to final volume at constant temperature, then cool by interaction with cold reservoir through diathermal wall. Let path 2 be the following: expand to final volume at constant pressure, then adjust temperature by interaction with heat reservoir through diathermal wall. Let path 3 be the following: cool to final temperature at constant volume, then expand to final pressure at constant temperature.

18-15. Derive the expression for the reversible isothermal work done on n moles of a gas by the surroundings at temperature T if the volume changes from V_1 to V_2 and the gas obeys the virial equation:

$$\frac{pV}{nRT} = 1 + \left(b - \frac{a}{RT}\right)\frac{n}{V}.$$

18-16. Calculate α and β for an ideal gas.

18-17. One hundred grams of nitrogen, N_2, treated as an ideal gas at 25°C and 725 mm pressure is expanded reversibly and isothermally to a volume of 700 l. What is the work in calories done on this system?

18-18. For which of the empirical equations (Eqs. 18-8 through 18-11) is the integration conveniently performed to find the work done on n moles in a reversible volume change from V_1 to V_2 at constant T? Integrate those that are not too difficult (van der Waals and Berthelot, at the very least).

18-19. One-half mole of an ideal gas is allowed to expand slowly while in contact with a heat bath at 400°K. The initial volume is 5.0 l, the final volume is 10 l. The work done is used to drive a refrigerator operating between a reservoir at 250°K and the one at 400°K. What is the maximum amount of heat withdrawn from the 250°K reservoir?

19 HEAT CAPACITIES

Section 18 was devoted to volumetric data on typical thermodynamic systems; this section discusses thermal data, namely, heat capacities. We first consider what it is thermodynamically that is learned when heat capacity data are obtained, then how to change from one kind of heat capacity data to another, and, finally, we consider typical values of heat capacities of various kinds of systems. Throughout this section we are presuming that only p-V work may be done, that the mole numbers are kept constant, and that no chemical reaction is occurring in the system.

DEFINITIONS OF HEAT CAPACITIES

The *heat capacity at constant volume*, C_V, is a property of a system defined as follows:

$$C_V \equiv \left(\frac{\partial U}{\partial T}\right)_V = T\left(\frac{\partial S}{\partial T}\right)_V, \qquad (19\text{-}1)$$

where the last identification was made from

$$dU = T\,dS - p\,dV \qquad (19\text{-}2)$$

by the process of "dividing by dT and imposing the constancy of V" (see Eqs. A-42 to 44 in Appendix A). We note that, if the temperature of a system is being changed through a heating process with only p-V work,

$$dU = dQ - p_{\text{ext}}\,dV. \qquad (19\text{-}3)$$

Then at constant volume,

$$dU = (dQ)_V \qquad \text{or} \qquad \boxed{\Delta U = Q_V.} \qquad (19\text{-}4)$$

Thus the heat absorbed in a constant-volume process is the same as ΔU if there is only p-V work. Therefore the ratio of the heat absorbed by the system at constant volume to the change in temperature is C_V:

$$C_V = \left(\frac{\partial U}{\partial T}\right)_V = \left(\frac{dQ}{dT}\right)_V \qquad (19\text{-}5)$$

Similarly, the **heat capacity at constant pressure**, C_p, is defined to be

$$\boxed{C_p \equiv \left(\frac{\partial H}{\partial T}\right)_p = T\left(\frac{\partial S}{\partial T}\right)_p,} \qquad (19\text{-}6)$$

where the last identification followed from

$$dH = T\,dS - V\,dp \qquad (19\text{-}7)$$

as above. From the definition of enthalpy,

$$H \equiv U + pV, \qquad (19\text{-}8)$$

$$dH = dU + p\,dV + V\,dp. \qquad (19\text{-}9)$$

Substitution for dU from Eq. 19-3 yields

$$dH = dQ + (p - p_{\text{ext}})\,dV + V\,dp. \qquad (19\text{-}10)$$

If the internal and external pressures are equal, then at constant pressure,

$$dH = (dQ)_p \qquad \text{or} \qquad \boxed{\Delta H = Q_p.} \qquad (19\text{-}11)$$

Thus the heat absorbed in a constant-pressure process is the same as ΔH if there is only p-V work and if the external and internal pressures are equal. Therefore the ratio of the heat absorbed by the system at constant pressure to the change in temperature is C_p:

$$C_p = \left(\frac{\partial H}{\partial T}\right)_p = \left(\frac{dQ}{dT}\right)_p. \qquad (19\text{-}12)$$

Comparing Eqs. 19-1 and 19-6 shows the effect of the two different processes considered.

The **specific heat** is the heat capacity for 1 g of material, usually for an experiment performed at constant pressure.

RELATIONSHIP BETWEEN C_p AND C_V

It sometimes happens that experimental data are available for either C_p or C_V, but that the other is desired. Finding the difference between the two is straightforward:

$$C_p - C_V = T\left(\frac{\partial S}{\partial T}\right)_p - T\left(\frac{\partial S}{\partial T}\right)_V. \qquad (19\text{-}13)$$

If Eq. A-33 in the form

$$\left(\frac{\partial S}{\partial T}\right)_p = \left(\frac{\partial S}{\partial T}\right)_V + \left(\frac{\partial S}{\partial V}\right)_T \left(\frac{\partial V}{\partial T}\right)_p \qquad (19\text{-}14)$$

is used for the first derivative in Eq. 19-13, this becomes

$$C_p - C_V = T\left(\frac{\partial S}{\partial V}\right)_T \left(\frac{\partial V}{\partial T}\right)_p = T\left(\frac{\partial p}{\partial T}\right)_V \left(\frac{\partial V}{\partial T}\right)_p, \qquad (19\text{-}15)$$

where the last step was from the Maxwell relation on the Helmholtz free energy, Eq. 13-34. The difference between the two heat capacities is obtainable simply from the two derivatives $(\partial p/\partial T)_V$ and $(\partial V/\partial T)_p$, both of which come immediately from the equation of state if it is known. Perhaps the commonest form in which Eq. 19-15 is written is

$$\boxed{C_p - C_V = \frac{TV\alpha^2}{\beta},} \qquad (19\text{-}16)$$

where α is the expansion coefficient and β the compressibility, as defined in Eqs. 18-22 and 18-23. Since β is always positive, C_p is always larger than C_V. One reason for this is that most materials expand when heated at constant pressure, thus doing work against the surroundings that they would not do at constant volume. Not only must enough heat be absorbed at constant pressure to raise the temperature but also enough energy must be absorbed to do this work of expansion. This intuitive argument is no good for those substances such as water just above its freezing point which contract on heating at constant pressure. These substances still have $C_p > C_V$, though for more complicated reasons based on the behavior of their constituent molecules.

The *ratio of the heat capacities* enters often into thermodynamic equations and is given the symbol γ:

$$\gamma \equiv \frac{C_p}{C_V} = 1 + \frac{TV\alpha^2}{C_V \beta}. \qquad (19\text{-}17)$$

It is always greater than unity.

PRESSURE AND VOLUME DEPENDENCE OF HEAT CAPACITIES

Suppose C_p is measured at a particular temperature and pressure, T and p_1, but one wants C_p at the same temperature but a different pressure p_2.

This is a typical problem of the type for which thermodynamics can give a ready answer. In Sec. 20, the general method of attack for these problems is considered. Here we simply compute the derivative $(\partial C_p/\partial p)_T$; its integration is undertaken in Sec. 20:

$$C_p = T\left(\frac{\partial S}{\partial T}\right)_p \qquad (19\text{-}18)$$

$$\left(\frac{\partial C_p}{\partial p}\right)_T = \left(\frac{\partial}{\partial p}\right)_T T\left(\frac{\partial S}{\partial T}\right)_p = T\left(\frac{\partial}{\partial p}\right)_T \left(\frac{\partial S}{\partial T}\right)_p, \qquad (19\text{-}19)$$

$$= T\left(\frac{\partial}{\partial T}\right)_p \left(\frac{\partial S}{\partial p}\right)_T, \qquad (19\text{-}20)$$

$$= -T\left(\frac{\partial}{\partial T}\right)_p \left(\frac{\partial V}{\partial T}\right)_p, \qquad (19\text{-}21)$$

$$\boxed{\left(\frac{\partial C_p}{\partial p}\right)_T = -T\left(\frac{\partial^2 V}{\partial T^2}\right)_p.} \qquad (19\text{-}22)$$

In Eq. 19-19, since T is constant in the $(\partial/\partial p)_T$, the factor T in Eq. 19-18 is just like any other constant and commutes with the differentiation. In Eq. 19-20, the order of differentiation was reversed. In Eq. 19-21, the Maxwell relation on the Gibbs function, Eq. 13-35, was used, and Eq. 19-22 is a simple rewriting of Eq. 19-21. The pressure dependence of C_p is shown by Eq. 19-22 to be obtainable from the derivative $(\partial^2 V/\partial T^2)_p$ that comes immediately from the equation of state.

Sometimes people are puzzled by the fact that the derivative on the left-hand side of Eq. 19-22 is taken at constant T while the two differentiations required on the right-hand side are at constant p and, in fact, the T is varying explicitly. This need cause no concern, however. Both C_p and V are functions of T and p. The quantity $(\partial C_p/\partial p)_T$ is another function of T and p, obtained from $C_p(T, p)$ by a particular recipe in which T is held constant. Similarly, $(\partial V/\partial T)_p$ and $(\partial^2 V/\partial T^2)_p$ are functions of T and p which come from $V(T, p)$ by a recipe in which p is held constant. It just happens that when $(\partial C_p/\partial p)_T$ is computed it must equal $-T(\partial^2 V/\partial T^2)_p$, that is, the two functions, regardless of the different ways in which they are computed, are equal. Thus if one can be found for a given system, the other is known.

A similar procedure is used in Prob. 19-2 to derive the related result for C_V:

$$\boxed{\left(\frac{\partial C_V}{\partial V}\right)_T = T\left(\frac{\partial^2 p}{\partial T^2}\right)_V.} \qquad (19\text{-}23)$$

EXPERIMENTAL HEAT CAPACITIES

Heat capacities of crystalline *solids* as functions of temperature start off at zero at $0°K$ and increase with increasing temperature.   Commonly, the temperature dependence is that predicted theoretically by Peter Debye in 1912, and shown graphically in Fig. 19-1.*   At low temperatures it obeys the so-called **Debye T^3 law**:

$$C_V = n\alpha T^3 \qquad \text{(solids, low temperature)}, \qquad (19\text{-}24)$$

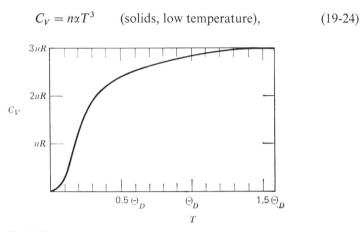

Fig. 19-1. Plot of C_V versus T for a typical crystalline solid.

where α is a constant for each compound which is fit by matching the right-hand side of Eq. 19-24 to observed values of C_V at low, but achievable, temperatures.   The entire C_V versus T curve, Fig. 19-1, may be obtained from the Debye theory.   The high-temperature plateau value of C_V is

$$C_V = 3nR \qquad \text{(solid elements, high temperature)} \qquad (19\text{-}25)$$

for solid elements.   This value of the heat capacity is often called the **law of Dulong and Petit**, because they proposed in 1819 that the molar heat capacities for all solid elements were the same.   It was L. Boltzmann in 1871 who derived the value $3nR$ from atomic theory.   Often, for *compounds*, the form of Dulong and Petit's law is valid, except instead of the number of moles n of the compound, one must substitute the number of gram-atoms nx, where x is the number of atoms per molecule:

$$C_V = 3nxR \qquad \text{(solid compounds, high temperature).} \qquad (19\text{-}26)$$

* F. C. Andrews, *Equilibrium Statistical Mechanics*, John Wiley & Sons, New York, 1963, section 29.

One further complication is that if there are electrons in the solid that are nearly free, as there are in metals, then they contribute a term

$$C_{\text{electron}} = n\gamma T \quad \text{(metals, low temperature)} \quad (19\text{-}27)$$

to the heat capacity.   Here, n is the number of moles of free electrons and γ is a constant that depends on the metal.*   The electronic contribution, Eq. 19-27, is small, and at room temperatures can be neglected compared to $3nR$.   It is observable only at low temperatures where $n\alpha T^3$ is very small.

Heat capacities for liquids, like all other properties of liquids, are much harder to derive from the molecular picture than for solids or gases.   When a solid melts to form a liquid, it usually undergoes a fairly small, discontinuous increase in heat capacity.   As the temperature of a liquid is increased, its heat capacity usually increases slowly.   When the liquid is vaporized, the gas usually has a considerably lower heat capacity than the liquid.   Of course, C_p for water is 1.00 cal/deg-g or 18.00 cal/deg-mole at 4.5°C, and not much different from that over a fairly wide temperature range.

The heat capacities for gases remain fairly constant, increasing only slightly as functions of temperature over quite wide ranges, from, say, 50 to 1000°K for monatomic, most diatomic, and many polyatomic gases.   These plateau values of C_V for dilute gases are given approximately by

$$C_V \approx \frac{v}{2} nR, \quad (19\text{-}28)$$

where $v = 3$ for monatomic gases, $v = 5$ for linear polyatomic gases (which include all diatomic gases plus gases like CO_2), and $v = 6$ for nonlinear polyatomic gases such as water.   As the temperature is increased still more, the heat capacity increases more rapidly.   The molecular theory permits accurate computation of heat capacities of dilute gases through use of statistical mechanics.†

Sometimes it is convenient to represent the temperature dependence of heat capacities by simple analytic formulas.   For most substances above room temperatures, the representation

$$C_p = n(a + bT + cT^{-2}) \quad (19\text{-}29)$$

seems to fit the data with great precision.   Here, a, b, and c are parameters chosen for each substance to fit best the available data.   These parameters

---

* J. E. Mayer and M. G. Mayer, *Statistical Mechanics*, John Wiley & Sons, New York, 1940, chapter 16.
† F. C. Andrews, *Equilibrium Statistical Mechanics*, John Wiley & Sons, New York, 1963, sections 18 and 19.

have been fit for a number of substances by Pitzer and Brewer.*   Other investigators have fit heat capacity data to the equation

$$C_p = n(a + bT + cT^2).$$ (19-30)

PROBLEMS

19-1. Derive Eq. 19-23 by a procedure similar to that used for obtaining Eq. 19-22.

19-2. A thermodynamics text states: "Addition of heat to a body whose parameters of state, apart from the temperature, are maintained constant always causes a rise of temperature, and therefore the principal specific heats of a body are always positive."   Analyze that statement.

19-3. Prove that $C_p - C_V = nR$ for an ideal gas.   If H_2 has $C_p = 6.8$ cal/deg-mole at 25°C, what is its value of C_V?

19-4. Compute $(\partial C_V/\partial V)_T$ for the gas of Prob. 18-15.

19-5. What is C_V (use calories) for monatomic gases such as He, Ne, Ar, etc.? What is C_p for them?   What is C_V for diatomic gases such as O_2, N_2, H_2, air, etc?   What is C_p for them?

19-6. What is Q for the process of cooling 1 kg of water vapor at 1 atm from 200 to 100°C?   What is Q for the process of cooling 1 kg of liquid water from 100 to 0°C?

19-7. How high must a waterfall be for its temperature at the bottom to be 0.1°C warmer than its temperature at the top?   If evaporation were taken into account, would the height be increased or decreased?

19-8. In the temperature range of 298 to 2000°K, Lewis, Randall, Pitzer, and Brewer (p. 66) give C_p of $NH_3(g)$ as $7.11 + 6.00 \times 10^{-3}T - 0.37 \times 10^5T^{-2}$ cal/deg-mole.   Find Q for the process of heating reversibly 30.0 g of NH_3 from 400 to 1300°K at 1 atm of pressure.

19-9. At 0°C, aluminum has the following properties: atomic weight $= 27.0$ g/mole, density $= 2.70$ g/cm³, $C_p = 0.220$ cal/g-deg, $\alpha = 71.4 \times 10^{-6}$ deg^{-1}, $\beta = 1.34 \times 10^{-12}$ cm²/dyne.   Calculate at 0°C the molar heat capacity at constant volume, and the ratio of heat capacities γ.

---

* G. N. Lewis, M. Randall, K. S. Pitzer, and L. Brewer, *Thermodynamics*, McGraw-Hill, Book Co., New York, 2d ed., 1961, p. 66.   They also give references to literature surveys of heat capacity data on p. 67.

20 CHANGE OF STATE PROBLEMS

A great many of the thermodynamics applications that one is likely to encounter are of the following type: A system in a given initial state undergoes a specified process to some final state. It is desired to find how various state functions change during the process so the final values of the functions may be calculated from the initial values, or vice versa. Also, for the process, one might want to determine the work or the heat. We already have treated elementary work and heat calculations, and from time time we will consider others. For work, the integral of $-p(V)\,dV$ is specified as soon as $p(V)$ is known for the path of the particular process under consideration. The heat is sometimes obtainable from a direct knowledge of the process; other times it must be obtained by difference from the change in internal energy and the work.

The next few sections treat a variety of processes, showing how the changes in equilibrium state variables can be related. The same approach is taken for all processes:

1. *The differential of the quantity whose change is sought is written in terms of the differentials of the most convenient state functions for the problem.* For example, if we want to know how V changes for a system of fixed composition in which T and p are being manipulated, we would write

$$V = V(T, p) \qquad (n_i \text{ constant}), \qquad (20\text{-}1)$$

$$dV = \left(\frac{\partial V}{\partial T}\right)_p dT + \left(\frac{\partial V}{\partial p}\right)_T dp \qquad (n_i \text{ constant}). \qquad (20\text{-}2)$$

2. *The partial derivatives occurring in the resulting expression are identified in terms of quantities whose values are either known or capable of approximation.* For Eq. 20-2, this could involve computing the values of the derivatives from the appropriate equation of state. Of course, if we knew the equation of state, we would not have to go through this whole process, since from it any one of the three—p, V, or T—may be found directly from the other two. More realistically, then, suppose only the values of α and β (Eqs, 18-22 and 18-23) are available from experiment, and not the equation of state. We then replace $(\partial V/\partial T)_p$ by $V\alpha$ and $(\partial V/\partial p)_T$ by $-V\beta$:

$$dV = V\alpha \, dT - V\beta \, dp \qquad (n_i \text{ constant}). \qquad (20\text{-}3)$$

3. *The equation is integrated along any path going from the initial to the final state along which the integrand is known.* Since the variables are state functions, any path chosen must lead to the same change in value. One simply picks the easiest path. For our example, to show how V changes with T and p, Eq. 20-3 rewritten

$$\frac{dV}{V} = d\,(\ln V) = \alpha \, dT - \beta \, dp \qquad (n_i \text{ constant}). \qquad (20\text{-}4)$$

may be integrated along any path from T_1 and p_1 to T_2 and p_2 for which α and β are known. If, for example, α is known as a function of T between T_1 and T_2 for the constant pressure p_1, and β is known as a function of p between p_1 and p_2 at the constant temperature T_2, then the path chosen for the integration of Eq. 10-4,

$$\int_1^2 d(\ln V) = \ln V_2 - \ln V_1 = \ln \frac{V_2}{V_1} = \int_{1 \text{ any path}}^2 (\alpha \, dT - \beta \, dp) \qquad (20\text{-}5)$$

is one going in two steps: $T_1, p_1 \to T_2, p_1$, at constant $p = p_1$; followed by $T_2, p_1 \to T_2, p_2$, at constant $T = T_2$. Along the first step, $dp = 0$ in the integrand of Eq. 20-5; along the second step, $dT = 0$. Then, for that path, Eq. 20-5 becomes

$$\ln \frac{V_2}{V_1} = \int_{T_1, p_1}^{T_2, p_1} \alpha(T, p_1) \, dT - \int_{T_2, p_1}^{T_2, p_2} \beta(T_2, p) \, dp. \qquad (20\text{-}6)$$

The values of the integrals can be found since the integrands are known. The result is a relationship among the six variables $V_1, T_1, p_1, V_2, T_2, p_2$, expressed in terms of quantities whose values are known. Any one of these variables could be obtained from knowing the other five.

Example 20-1. Compute $C_p(T, p_2)$ knowing $C_p(T, p_1)$ and necessary volumetric data.

Solution: The first step is to express the total differential of C_p in terms of changes in the independent variables T and p most appropriate to the available data:

$$C_p = C_p(T, p), \qquad (20\text{-}7)$$

$$dC_p = \left(\frac{\partial C_p}{\partial T}\right)_p dT + \left(\frac{\partial C_p}{\partial p}\right)_T dp. \qquad (20\text{-}8)$$

The second step is to identify the derivatives. The first, $(\partial C_p/\partial T)_p$, can be obtained usually only by measuring C_p as a function of temperature. However, the problem at hand permits us to forget that derivative since the temperature is not changed.   In Sec. 19 the constant-temperature derivatives of the heat capacities were related to volumetric data, so substitution from Eq. 19-22 yields

$$dC_p = -T \left(\frac{\partial^2 V}{\partial T^2}\right)_p dp \qquad (T \text{ constant}). \qquad (20\text{-}9)$$

The third step is simply to integrate this:

$$C_p(T, p_2) = C_p(T, p_1) - T \int_{T, p_1}^{T, p_2} \left(\frac{\partial^2 V}{\partial T^2}\right)_p dp. \qquad (20\text{-}10)$$

This shows how the derivatives found in Eqs. 19-22 and 19-23 are actually used and is a good example of the strength of thermodynamics.   Volumetric measurements, relatively easily made, are used to find heat capacities (which are relatively hard to measure) at all accessible pressures if a single measured value of the heat capacity is available for the temperature of interest.

For C_V the formula analogous to Eq. 20-10 is

$$C_V(T, V_2) = C_V(T, V_1) + T \int_{T, V_1}^{T, V_2} \left(\frac{\partial^2 p}{\partial T^2}\right)_V dV. \qquad (20\text{-}11)$$

PROBLEMS

20-1. Use the values of α and β for ideal gases in Eq. 20-6 to find the relationship among $p_1, V_1, T_1, p_2, V_2, T_2$ and compare the result with the ideal-gas law to see if it makes sense.

20-2. Substitute the appropriate derivatives into Eq. 20-2 for a gas which obeys the equation of state $p(V - nb) = nRT$, where b is a constant.   Integrate to find $V_2 - V_1$ for a process going from V_1, p_1, T_1 to V_2, p_2, T_2, for each of the following paths: (a) $p_1, T_1 \rightarrow p_1, T_2$, and then $p_1, T_2 \rightarrow p_2, T_2$.   (b) $p_1, T_1 \rightarrow p_2, T_1$, and then $p_2, T_1 \rightarrow p_2, T_2$.   The fact that the answers are the same is reassuring, since ΔV must certainly be independent of the path.

20-3. A gas obeying Berthelot's equation has heat capacity $C_V(T, V_0)$; compute C_V at the different volume V_1 but at the same temperature T.

21 | ENTROPY CHANGES

NO PHASE TRANSITION

In Sec. 20, a general approach was presented for relating initial and final values of system properties for changes in state. In this section the method is used to compute entropy changes for systems of constant composition; systems of variable composition are treated in Secs. 26 to 30. First, suppose our knowledge of states 1 and 2 is explicitly of T_1, V_1, and T_2, V_2. Then the differential of S is written in terms of dT and dV:

$$S = S(T, V) \qquad (n_i \text{ constant}), \tag{21-1}$$

$$dS = \left(\frac{\partial S}{\partial T}\right)_V dT + \left(\frac{\partial S}{\partial V}\right)_T dV \qquad (n_i \text{ constant}). \tag{21-2}$$

The derivative $(\partial S/\partial T)_V$ is identified from the definition, Eq. 19-1, of C_V:

$$C_V = T\left(\frac{\partial S}{\partial T}\right)_V \tag{21-3}$$

and the derivative $(\partial S/\partial V)_T$ is obtained from the Maxwell relation, Eq. 13-34 on the Helmholtz free energy:

$$\left(\frac{\partial S}{\partial V}\right)_T = \left(\frac{\partial p}{\partial T}\right)_V. \tag{21-4}$$

Thus, Eq. 21-2 for dS becomes the following, written now in terms of quantities directly accessible to measurement:

$$\boxed{dS = \frac{C_V}{T} dT + \left(\frac{\partial p}{\partial T}\right)_V dV.} \tag{21-5}$$

This can be integrated over any path between states 1 and 2 for which the integrand is known. One needs to know C_V between the lower and upper temperatures, and one needs to know $(\partial p/\partial T)_V$. If one knew α and β instead of $(\partial p/\partial T)_V$, it is easy to show (see Prob. 21-1 and the definitions, Eqs. 18-22 and 18-23) that

$$\left(\frac{\partial S}{\partial V}\right)_T = \left(\frac{\partial p}{\partial T}\right)_V = \frac{\alpha}{\beta}. \tag{21-6}$$

We are free to integrate Eq. 21-5 over any path coupling states 1 and 2 for which the integrand is known.   For example, if C_V was known as a function of T for $V = V_1$ and $(\partial p/\partial T)_V$ was known as a function of V for $T = T_2$, then the path $T_1, V_1 \to T_2, V_1 \to T_2, V_2$ would be convenient:

$$S_2 - S_1 = \Delta S = \int_{T_1, V_1}^{T_2, V_1} \frac{C_V}{T} \, dT + \int_{T_2, V_1}^{T_2, V_2} \left(\frac{\partial p}{\partial T}\right)_V dV. \tag{21-7}$$

This path is sketched in Fig. 21-1.   On the other hand, C_V might be known as

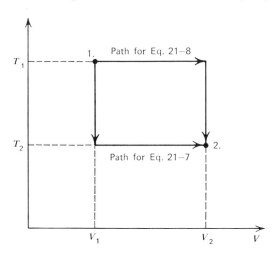

Fig. 21-1. Paths of integration of Eq. 21-5 yielding Eqs. 21-7 and 21-8.

a function of T for $V = V_2$ and $(\partial p/\partial T)_V$ known as a function of V for $T = T_1$. In that case, the path $T_1, V_1 \to T_1, V_2 \to T_2, V_2$ would be convenient:

$$S_2 - S_1 = \Delta S = \int_{T_1, V_2}^{T_2, V_2} \frac{C_V}{T} \, dT + \int_{T_1, V_1}^{T_1, V_2} \left(\frac{\partial p}{\partial T}\right)_V dV. \tag{21-8}$$

This path also is shown in Fig. 21-1.   The two integration paths must give the same result.

Now suppose our knowledge of states 1 and 2 is explicitly of T_1, p_1, and T_2, p_2, rather than of T's and V's.   Here it is more convenient to start with

$$S = S(T, p) \qquad (n_i \text{ fixed}), \tag{21-9}$$

rather than with Eq. 21-1. Methods completely analogous to the derivation of Eq. 21-5 yield (see Prob. 21-2)

$$dS = \frac{C_p}{T} dT - \left(\frac{\partial V}{\partial T}\right)_p dp = \frac{C_p}{T} dT - V\alpha \, dp, \qquad (21\text{-}10)$$

which may be integrated over any path connecting states 1 and 2.

If the wrong heat capacity data should be available (that is, C_p is known when C_V is called for, or vice versa, or $C_V(T, V)$ is given for the wrong volume or C_p for the wrong pressure), then Eqs. 19-16, 19-22, or 19-23 may be used to relate the quantities needed to the quantities known. We thus can conclude that given two states of the same composition and not separated by a phase transition, $S_2 - S_1$ may be found from knowledge of (a) one heat capacity or the other as a function of T for some path between T_1 and T_2, and (b) appropriate volumetric data for the p-V-T range of interest.

One last complication also arises, namely, some systems or heat reservoirs, such as the atmosphere or a lake, are so big that the integration of the term $C_p \, dT/T$ or $C_V \, dT/T$ is impossible. The temperature of the reservoir remains nearly constant. In this case, one pulls the T outside the integral since it is constant, and gets

$$\Delta S = \int \frac{C_V \, dT}{T} = \frac{1}{T} \int C_V \, dT = \frac{\Delta U}{T} \qquad (V \text{ and } T \text{ constant}), \quad (21\text{-}11)$$

$$\Delta S = \int \frac{C_p \, dT}{T} = \frac{1}{T} \int C_p \, dT = \frac{\Delta H}{T} \qquad (p \text{ and } T \text{ constant}). \quad (21\text{-}12)$$

The expressions are correct regardless of whether the process is reversible or irreversible, since entropy is a function of state.

Example 21-1. Find ΔS for the process of taking n moles of an ideal gas with constant heat capacity from T_1 and V_1 to T_2 and V_2.

Solution: Rather than memorizing anything, one would normally proceed to derive Eq. 21-5 directly from Eq. 21-1. One then finds $(\partial p/\partial T)_V$ for an ideal gas:

$$p = \frac{nRT}{V}, \qquad (21\text{-}13)$$

$$\left(\frac{\partial p}{\partial T}\right)_V = \frac{nR}{V}, \qquad (21\text{-}14)$$

$$dS = \frac{C_V}{T} dT + \frac{nR}{V} dV \qquad (\text{ideal gas}). \qquad (21\text{-}15)$$

Since C_V is constant, it makes no difference what path is chosen for the integration of Eq. 21-15; the result is

$$\Delta S = C_V \ln \frac{T_2}{T_1} + nR \ln \frac{V_2}{V_1} \quad \text{(ideal gas, } C_V \text{ constant).} \quad (21\text{-}16)$$

CHANGE IN PHASE

If states 1 and 2 refer to the material in difference phases (for example, solid and liquid, solid and vapor, liquid and vapor), the state functions like volume, energy, and entropy depend not just on T and p, but they are also functions of which phase one has.   For example, the transition from ice to water may occur at the constant $T = 0°C$ and $p = 1$ atm, but the molar volume, energy, entropy, etc., change during the transition.

There are two equivalent derivations of the expression of the entropy change for a phase transition.   The first is simply to integrate the equation defining entropy,

$$dS = \frac{dQ_{\text{rev}}}{T} \quad (21\text{-}17)$$

for the constant-temperature phase transition,

$$S_{\text{trans}} = S(T, p, \text{phase 2}) - S(T, p, \text{phase 1})$$

$$= \int_1^2 dS = \int_1^2 \frac{dQ_{\text{rev}}}{T} = \frac{1}{T} \int_1^2 dQ_{\text{rev}} = \frac{Q_{\text{rev}}}{T}. \quad (21\text{-}18)$$

Now, Eq. 19-11 may be used to identify the heat absorbed in this constant-pressure process as the enthalpy change

$$Q_{\text{rev}} = \Delta H_{\text{trans}} \equiv H(T, p, \text{phase 2}) - H(T, p, \text{phase 1}). \quad (21\text{-}19)$$

Thus

$$\boxed{\Delta S_{\text{trans}} = \frac{\Delta H_{\text{trans}}}{T}} \quad \text{(phase transition).} \quad (21\text{-}20)$$

For the melting of a solid, this is written $\Delta S_{\text{fus}} = \Delta H_{\text{fus}}/T$, where fus stands for fusion.   For vaporization of a liquid, this is written $\Delta S_{\text{vap}} = \Delta H_{\text{vap}}/T$, and for sublimation, $\Delta S_{\text{sub}} = \Delta H_{\text{sub}}/T$.   Terms of the form of Eq. 21-20 must be added to all equations like Eqs. 21-7 or 21-8 for every change of phase which occurs along the path of integration.

A second derivation of this same result is based on the conclusion reached from Eq. 15-14 that matter of type j will flow from regions of high chemical potential to low (and thereby increase the entropy) until at equilibrium μ_j has the same value in all parts of the system accessible to j. During the phase transition, there is an equilibrium between the material in phase 1 and phase 2:

$$\mu^{(1)} = \mu^{(2)} \qquad \text{(phase equilibrium).} \tag{21-21}$$

Here we can use the result, Eq. 12-20, that $\mu = u - Ts + pv$, or $\mu = h - Ts$, and this becomes

$$h^{(1)} - Ts^{(1)} = h^{(2)} - Ts^{(2)}, \tag{21-22}$$

$$T(s^{(2)} - s^{(1)}) = h^{(2)} - h^{(1)}, \tag{21-23}$$

$$\Delta s_{\text{trans}} = \frac{\Delta h_{\text{trans}}}{T} \qquad \text{(phase transition).} \tag{21-24}$$

The result is analogous to Eq. 21-12, obtained previously.

LOW TEMPERATURES

The usual reference state for entropy values is the pure crystal at $0°K$. This poses an experimental problem, since $0°K$ cannot be reached.* However, we can usually trust the Debye T^3 law for heat capacities for temperatures from $0°K$ up to some experimentally reachable low temperature, say, T_{exp}. Values of T_{exp} in the range of 4 to $12°K$ are not too hard or expensive to reach. We can compute S at T_{exp}:

$$S(T_{\text{exp}}) = S(0°K) + \int_{0°}^{T_{\text{exp}}} \frac{C\,dT}{T}. \tag{21-25}$$

We have not specified C_p or C_V, since at low temperatures the difference $TV\alpha^2/\beta$ between them is small. If now we use the Debye law,

$$C = n\alpha T^3, \tag{21-26}$$

$$S(T_{\text{exp}}) = S(0°K) + \int_{0°}^{T_{\text{exp}}} \frac{n\alpha T^3}{T}\,dT \tag{21-27}$$

$$= S(0°K) + \tfrac{1}{3}n\alpha T_{\text{exp}}^3 \tag{21-28}$$

$$S(T_{\text{exp}}) = S(0°K) + \tfrac{1}{3}C(T_{\text{exp}}) \tag{21-29}$$

---

* This fact is a consequence of the so-called third law of thermodynamics, which is discussed in Sec. 30.

Thus relative to its value at $0°K$, the entropy at any temperature up to which the Debye formula applies is just one-third the heat capacity at that temperature, something which can, of course, be measured directly.

SUMMARY

The entropy might be written, then, for a system of fixed composition relative to its entropy at $0°K$ as follows:

$$S(T, p) = S(0°K) + \tfrac{1}{3}C(T_{exp}) + \int_{T_{exp}, p_0}^{T, p_0} \frac{C_p(T, p_0)}{T} dT$$

$$- \int_{p_0, T}^{p, T} \left(\frac{\partial V}{\partial T}\right)_p dp + \sum_{\substack{\text{phase} \\ \text{trans}}} \frac{\Delta H_{trans}}{T_{trans}} \qquad (n_i \text{ constant}). \quad (21\text{-}30).$$

The question of how to assign the reference value $S(0°K)$ to the entropy for various substances is treated in Sec. 30 in the discussion of the third law of thermodynamics.   Usually, $S(0°K)$ may be set equal to zero.

Example 21-2. Forty grams of water is heated from 25 to $100°C$ and then vaporized.   The heat of vaporization of water at $100°C$ is 539.7 cal/g. Calculate ΔS.

Solution: For the step in which the water is heated,

$$\Delta S_{heat} = \int_{T_1}^{T_2} \frac{C_p}{T} dT = C_p \ln \frac{T_2}{T_1}, \qquad (21\text{-}31)$$

$$\Delta S_{heat} = \frac{(40 \text{ g})(1.00 \text{ cal})(2.303)}{(\text{deg-g})} \log \frac{373}{298}, \qquad (21\text{-}32)$$

$$\Delta S_{heat} = 8.92 \text{cal/deg}. \qquad (21\text{-}33)$$

For the vaporization,

$$\Delta S_{vap} = \frac{\Delta H_{vap}}{T_{vap}} = \frac{(40 \text{ g})(539.7 \text{ cal})}{(\text{g})(373 \text{ deg})} = 57.9 \text{ cal/deg}, \qquad (21\text{-}34)$$

and

$$\Delta S_{total} = \Delta S_{heat} + \Delta S_{vap} = 66.8 \text{cal/deg}. \qquad (21\text{-}35)$$

The answer could also have been given in so-called **entropy units**, abbreviated eu, where *the entropy unit is defined to be 1 cal/deg.*

PROBLEMS

21-1. Prove Eq. 21-6.

21-2. Prove Eq. 21-10.

21-3. Find ΔS for heating 100 ml of $H_2O(l)$ from 0 to 100°C at 1 atm of pressure.

21-4. Find ΔS for heating 10.0 g of helium gas (treated as ideal) from 20 to 700°C at constant volume at 50 l.

21-5. Find ΔS for cooling 20.0 g of oxygen gas (treated as ideal) from 500 to 100°C at constant pressure of 730 mm.

21-6. Find ΔS for the process of taking n-moles of an ideal gas with constant heat capacity from T_1 and p_1 to T_2 and p_2.

21-7. Calculate the entropy changes for the following processes: (*a*) Melting of 1 mole of argon at its melting point, -190°C. The molar enthalpy of fusion of argon is 6.71 cal/g. (*b*) Evaporation of 32.0 g of liquid oxygen at its boiling point, -182.97°C. The molar enthalpy of vaporization is 1.630 kcal/mole. (*c*) The heating of 10.0 g of hydrogen sulfide from 50 to 100°C. The heat capacity of H_2S in cal/deg-mole at constant pressure is $7.15 + 0.00332T$, where T is in °K.

21-8. One kilogram of steam is condensed at 100°C, and the water is cooled to 0°C and frozen to ice. What is the entropy change? What is the enthalpy change? The heat of vaporization at the boiling point and the heat of fusion at the freezing point are 539.7 and 79.7 cal/g respectively.

21-9. Calculate the change in entropy if 400 g of lead at -10°C is placed in contact with 700 g of lead at 100°C and the combination is allowed to reach equilibrium, thermally isolated from the surroundings. The specific heat of lead is 0.03 cal/g-deg.

21-10. Eliminate dT between Eqs. 21-5 and 21-10 to prove the following for the relation between dS, dV, and dp:

$$dS = \frac{\gamma}{\gamma - 1}\left(\frac{\partial p}{\partial T}\right)_V dV + \frac{1}{\gamma - 1}\left(\frac{\partial V}{\partial T}\right)_p dp, \qquad (21\text{-}36)$$

where γ is the ratio of the heat capacities. This too can be integrated along any path between state 1 and state 2 for which the integrand is known.

21-11. (*a*) One kilogram of water is heated reversibly from 27 to 87°C at 1 atm (by use of an infinite series of heat reservoirs). What is ΔS for the water, for the reservoirs, and for the composite (water plus reservoirs)? (*b*) If the water were placed directly into a very large reservoir at 87°C, what is ΔS for the water, for the reservoir, and for the composite?

21-12. An electric current of 10 amp flows through a resistor of 20 ohms which is kept at the constant temperature of 10°C by running water. In 1 sec, what is the entropy change of the resistor and the entropy change of the water?

21-13. An electric current of 10 amp flows through a thermally insulated resistor of 20 ohms, initially at a temperature of 10°C, for 1 sec. If the resistor has a mass of 5 g and $C_p = 0.20$ cal/g-deg, what is the entropy change of the resistor and the entropy change of its surroundings?

21-14. Sufficient steam at 100°C and 1 atm of pressure is condensed into 100 g of water at 50°C to raise its temperature just to 100°C. Calculate the total entropy change. The heat of vaporization can be taken as 540 cal/g and the heat capacity of water as 1.00 cal/deg-g.

21-15. *Trouton's rule* states that so-called normal or nonpolar liquids have ΔS

of vaporization at their normal boiling points of about 21 eu/mole. Benzene boils at 80°C; estimate its heat of vaporization.

21-16. Given two objects of constant heat capacities C_1 and C_2 initially at T_1 and T_2. If they are simply placed in thermal contact, zero work is done on the surroundings, and (a) the final temperature is what? If a reversible heat engine is placed between them, harnessing their approach to equilibrium to do maximum work, (b) what is the final temperature? (c) How much work is done on the surroundings?

21-17. Do Prob. 21-16, replacing object 2 by an essentially infinite heat reservoir, such as a lake or the atmosphere.

21-18. The usual ΔS of fusion is given at constant pressure. Express ΔS of fusion at constant volume in terms of the value at constant pressure, α and β for the liquid (presumed constant), and the molar volumes v_l and v_s of liquid and solid.

21-19. If $C_p = na + nbT + ncT^2$ for a substance, (a) what is Q for the reversible cooling from T_1 to T_2 at fixed pressure, and (b) what is ΔS for that process?

21-20. A diathermal wall separates two heat reservoirs maintained at 20 and 100°C respectively. In the steady state, 200 cal of heat is conducted every minute through the wall and the properties of the wall are not changing with time. What is the entropy change of the wall itself per minute in the steady state? What is the total entropy change per minute resulting from the irreversible process occurring in the wall?

21-21. A mole of supercooled water in contact with a large heat reservoir at $-10°C$ freezes isothermally at $-10°C$. What is the difference between ΔS for this and ΔS for the freezing at 0°C? $C_p(\text{ice}) = 0.5$ cal/deg-g; $C_p(\text{water}) = 1.0$ cal/deg-g; heat of fusion of ice = 80 cal/g at 0°C.

22   ISENTROPIC PROCESSES

Systems which are thermally isolated from the rest of the world and have only reversible processes occurring within them have constant entropy (that is, are *isentropic*). Of course, neither perfect reversibility nor perfect thermal isolation is possible to achieve, but for many processes the system can very accurately be treated as isentropic. For example, consider a gas contained in a cylinder by a piston. If the piston and cylinder walls are poor conductors of heat, the piston could undergo several cycles up and down before any appreciable amount of heat interaction occurred with the gas. If the piston motion were so fast as to generate waves in the gas, the irreversible damping out of these waves would destroy the constancy of S for the process. But if the piston moved slowly enough so that this effect was negligible, then even if there was considerable friction between the piston and the cylinder walls, the gas would remain at almost constant entropy. True, work is lost through the friction which heats up the piston itself and the walls of the cylinder, and if the rubbing continues long enough, a significant heat interaction with the gas must occur. But if the gas is studied over the course of only one or two strokes of the piston, the near constancy of the entropy is the best starting place for analysis.

There are many other processes which similarly meet the requirements for constant entropy, namely, they occur slowly with respect to the internal relaxation times of the system and they have a negligible heat interaction with the surroundings.

For simple systems of fixed composition, any two properties such as T and V are enough to specify the equilibrium state. The set of all pairs of the variables T and V which have the same value of S would form a line in the T-V plane. The relationship among points T_1, V_1 and T_2, V_2 which have the same entropy may be found by integrating Eq. 21-5 along any path for which the integrand is known, and then setting $\Delta S = 0$. If the path T_1, $V_1 \to T_2$, $V_1 \to T_2$, V_2, shown in Fig. 21-1 for Eq. 21-7, is convenient, one gets

$$\int_{T_1, V_1}^{T_2, V_1} \frac{C_V}{T} dT + \int_{T_2, V_1}^{T_2, V_2} \left(\frac{\partial p}{\partial T}\right)_V dV = 0 \qquad (S \text{ constant}). \qquad (22\text{-}1)$$

This is an equation from which the relationship among T_1, V_1, T_2, V_2 may be found for any isentropic process. If the path T_1, $V_1 \to T_1$, $V_2 \to T_2$, V_2, shown in Fig. 21-1 for Eq. 21-8, is used instead, one gets the analogous expression, which also is correct:

$$\int_{T_1, V_2}^{T_2, V_2} \frac{C_V}{T} dT + \int_{T_1, V_1}^{T_1, V_2} \left(\frac{\partial p}{\partial T}\right)_V dV = 0 \qquad (S \text{ constant}). \qquad (22\text{-}2)$$

A similar approach may be used to relate the initial and final temperatures and pressures for isentropic processes by integrating Eq. 21-10 and setting $\Delta S = 0$:

$$\int_{T_1, p_1}^{T_2, p_1} \frac{C_p}{T} dT - \int_{T_2, p_1}^{T_2, p_2} \left(\frac{\partial V}{\partial T}\right)_p dp = 0 \qquad (S \text{ constant}). \qquad (22\text{-}3)$$

The path T_1, $p_1 \to T_1$, $p_2 \to T_2$, p_2 could just as well have been chosen. The initial and final pressures and volumes may be related by integrating Eq. 21-36 (see Prob. 21-10):

$$\int_{V_1, p_1}^{V_2, p_1} \frac{\gamma}{\gamma - 1} \left(\frac{\partial p}{\partial T}\right)_V dV + \int_{V_2, p_1}^{V_2, p_2} \frac{1}{\gamma - 1} \left(\frac{\partial V}{\partial T}\right)_p dp = 0 \qquad (S \text{ constant}),$$

$$(22\text{-}4)$$

where γ is the ratio of the heat capacities, C_p/C_V. Here again, the alternate path where p is changed first and then V is changed could have been used, if desired. If γ is approximately a constant in the range of interest:

$$\gamma \int_{V_1, p_1}^{V_2, p_1} \left(\frac{\partial p}{\partial T}\right)_V dV + \int_{V_2, p_1}^{V_2, p_2} \left(\frac{\partial V}{\partial T}\right)_p dp = 0 \qquad (S \text{ and } \gamma \text{ constant}). \quad (22\text{-}5)$$

These equations are completely general and can be used for any equation of state and any functional dependence of the heat capacity. Clearly they are impossible to remember, but the appropriate one can be derived when needed after a little practice.

IDEAL GASES

The ideal-gas law is the easiest to work with, and results obtained with it give insight into the behavior of real gases. If the ideal-gas law is used to find the derivatives in Eq. 22-5, one obtains

$$\gamma \int_{V_1, p_1}^{V_2, p_1} \frac{nR \, dV}{V} + \int_{V_2, p_1}^{V_2, p_2} \frac{nR \, dp}{p} = 0, \qquad (22\text{-}6)$$

$$\gamma \ln \frac{V_2}{V_1} + \ln \frac{p_2}{p_1} = 0, \qquad (22\text{-}7)$$

$$\ln \left(\frac{V_2}{V_1}\right)^\gamma = \ln \frac{p_1}{p_2}, \qquad (22\text{-}8)$$

$$\left(\frac{V_2}{V_1}\right)^\gamma = \frac{p_1}{p_2} \qquad \text{or} \qquad p_1 V_1^\gamma = p_2 V_2^\gamma, \qquad (22\text{-}9)$$

or

$$\boxed{pV^\gamma = \text{const.}} \qquad (S \text{ and } \gamma \text{ constant, ideal gas}). \qquad (22\text{-}10)$$

In Eq. 22-9, recognition was made of the fact that if $\ln a = \ln b$, then $a = b$. The result, Eq. 22-10, is one of those few equations that is perhaps best remembered, provided the qualifications of constant γ and ideal gas are also remembered.

If one wants an isentropic equation for ideal gases which involves temperature, one may simply replace the appropriate variable in the expression $pV^\gamma = $ constant by its equivalent, using $pV = nRT$:

$$pV^\gamma = \text{const.}, \qquad p = \frac{nRT}{V} \qquad (22\text{-}11)$$

$$TV^{\gamma-1} = \text{const.} \quad (S \text{ and } \gamma \text{ constant, ideal gas}), \qquad (22\text{-}12)$$

where, of course, the constant of Eq. 22-12 differs from that of Eq. 22-10. Similarly,

$$p V^\gamma = \text{const.,} \qquad V = \frac{nRT}{p} \qquad (22\text{-}13)$$

$$p^{1-\gamma} T^\gamma = \text{const.} \quad (S \text{ and } \gamma \text{ constant, ideal gas).} \qquad (22\text{-}14)$$

Since for ideal gases $C_p - C_V = nR$, other forms equivalent to Eqs. 22-12 and 22-14, derived below in Probs. 22-1 and 22-2, are the following:

$$T V^{nR/C_V} = \text{const.} \quad (S \text{ and } \gamma \text{ constant, ideal gas),} \qquad (22\text{-}15)$$

$$T^{-C_p/nR} p = \text{const.} \quad (S \text{ and } \gamma \text{ constant, ideal gas).} \qquad (22\text{-}16)$$

Example 22-1. In Sec. 11 it was noted that a Carnot cycle is reversible and has heat interactions only with reservoirs at T_1 and T_2; thus the steps in the cycle in which the temperature changes must be adiabatic (therefore isentropic). In Fig. 22-1, two isotherms and two isentropes have been drawn

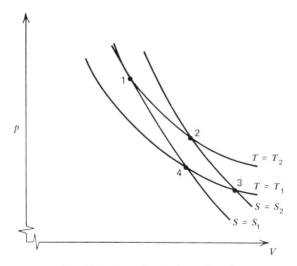

Fig. 22-1. Carnot cycle for p-V work.

in the p-V plane for a typical substance. The resulting closed figure is a Carnot cycle, which may be traversed either clockwise or counterclockwise. Prove that the slope of the isentrope through an arbitrary point, such as point 2 in Fig. 22-1, is γ times the slope of the isotherm through that point.

Solution: Solve Eq. 21-36 for dp at constant S; then "division by dV" (see Eqs. A-41 to 44) yields

$$\left(\frac{\partial p}{\partial V}\right)_S = \frac{-\gamma \left(\frac{\partial p}{\partial T}\right)_V}{\left(\frac{\partial V}{\partial T}\right)_p} = \gamma \left(\frac{\partial p}{\partial V}\right)_T. \qquad (22\text{-}17)$$

The last step was made by using Eq. A-22.

An analysis of Eq. 22-17 is worthwhile. The value of $(\partial p/\partial V)_T$ was discussed in relation to Fig. 18-1 for various states of matter: very large and negative for condensed phases (solid and liquid), much smaller but still negative for gases, tending toward zero as $V \to \infty$. Since γ is always greater than unity, the absolute value of $(\partial p/\partial V)_S$ is always larger than that of $(\partial p/\partial V)_T$, though both are negative. Therefore, the typical Carnot cycle of Fig. 22-1 may be drawn with confidence in the fact that the isentropes are steeper than the isotherms.

PROBLEMS

22-1. Derive Eq. 22-15 from Eq. 22-10.
22-2. Derive Eq. 22-16 from Eq. 22-10.
22-3. If an ideal monatomic gas doubles its volume in a reversible adiabatic process, how is the temperature changed?
22-4. How is the pressure changed in Prob. 22-3?
22-5. If an ideal diatomic gas doubles its volume in a reversible adiabatic process, how is the temperature changed?
22-6. Five moles of an ideal gas, $C_p = 7n$ in units of cal/deg, expand reversibly and adiabatically from 10 atm and 127°C to a final pressure of 2 atm. What is the final temperature? The final volume?
22-7. Calculate the work W done on n moles of an ideal gas with constant γ, which are expanded reversibly and adiabatically from V_1 to V_2; the initial temperature is T_1.
22-8. Two moles of an ideal diatomic gas initially at 30°C are expanded reversibly from 25 to 50 l, (a) adjusting the temperature so the pressure is constant, (b) adiabatically. Calculate W in both of the cases.
22-9. Several thermodynamics texts state that p_1, V_1, p_2, V_2 may be related for an isentropic process simply by integrating Eq. 22-17:

$$\int_{p_1}^{p_2} dp = \int_{V_1}^{V_2} \gamma \left(\frac{\partial p}{\partial V}\right)_T dV \qquad (S \text{ constant}), \qquad (22\text{-}18)$$

rather than using Eq. 22-4. However, in most practical cases, Eq. 22-18 is useless. Why?
22-10. The *velocity of sound* c in a gas is given by $c^2 = (\partial p/\partial \rho)_s$, where ρ is the mass per unit volume. (a) Prove that $c^2 = \gamma RT/M$ for an ideal gas, where M is the molecular weight. (b) Find c^2 for a van der Waals gas.
22-11. Suppose a gas is described by Eq. 18-21, truncated after the second virial term, with $B'(T)$ given by Eq. 18-20. If $C_p(p_1)$ is known and is independent of T, relate T_1, p_1, T_2, p_2 for an isentropic process.

22-12. The *adiabatic compressibility* β_s is defined like the isothermal compressibility β, except entropy rather than temperature is held constant. Calculate β_s for the aluminum of Prob. 19-9.

22-13. A pressure jump apparatus expands water from 61 to 1 atm adiabatically and almost reversibly. Estimate the temperature change for room temperature water, deriving what equations are needed; $\alpha = 2.1 \times 10^{-4}$ deg^{-1}.

22-14. Prove $(\partial T/\partial V)_s = -(T/C_V)(\partial p/\partial T)_V$.

23 ENERGY CHANGES AND JOULE EXPANSIONS

Suppose one wanted to relate initial and final values of the internal energy to initial and final temperatures and volumes for systems of fixed composition. The procedure is by now automatic:

$$U = U(T, V) \qquad (n_i \text{ constant}), \tag{23-1}$$

$$dU = \left(\frac{\partial U}{\partial T}\right)_V dT + \left(\frac{\partial U}{\partial V}\right)_T dV \qquad (n_i \text{ constant}). \tag{23-2}$$

The derivative $(\partial U/\partial T)_V$ is immediately recognized:

$$\left(\frac{\partial U}{\partial T}\right)_V = C_V, \tag{23-3}$$

and $(\partial U/\partial V)_T$ is identified from the Gibbs equation for dU:

$$dU = T\,dS - p\,dV. \tag{23-4}$$

Thus

$$\left(\frac{\partial U}{\partial V}\right)_T = T\left(\frac{\partial S}{\partial V}\right)_T - p = T\left(\frac{\partial p}{\partial T}\right)_V - p, \tag{23-5}$$

where the final form followed by Maxwellization. This result was previously obtained in Eqs. 14-1, 14-2, and 14-3. The use of Eqs. 23-3 and 23-5 in Eq. 23-2 yields

$$dU = C_V\,dT + \left[T\left(\frac{\partial p}{\partial T}\right)_V - p\right] dV. \tag{23-6}$$

The difference between Eqs. 23-4 and 23-6 is interesting. The form of Eq. 23-4 is simple and contains no derivatives because S and V are the natural variables with which to express U as a thermodynamic potential. When use of S as an independent variable is replaced by T, the different and more complicated form, Eq. 23-6, results.

The result, Eq. 23-6, may be integrated over any path between states 1 and 2 for which the integrand is known. For example, if the path $T_1, V_1 \to T_2, V_1 \to T_2, V_2$ is chosen.

$$U_2 - U_1 = \int_{T_1, V_1}^{T_2, V_1} C_V \, dT + \int_{T_2, V_1}^{T_2, V_2} \left[T\left(\frac{\partial p}{\partial T}\right)_V - p \right] dV. \qquad (23\text{-}7)$$

Other paths, for example, $T_1, V_1 \to T_1, V_2 \to T_2, V_2$, could be used as well.

There is considerable utility in being able to find ΔU for a process. One use is that by knowing ΔU for states 1 and 2 and knowing $W = -\int p_{\text{ext}}(V) \, dV$ for the path taken in a particular process, the heat Q absorbed by the system may be found by difference from the first law: $\Delta U = Q + W$.

Another use of these results is that they permit an analysis of the experiment of the two bottles discussed in Sec. 8 and illustrated in Fig. 23-1. This is

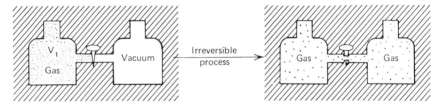

Fig. 23-1. The Joule expansion of a gas.

the so-called **Joule experiment**, an irreversible expansion of a gas into a vacuum. The system goes from one equilibrium state to another as a result of the releasing of the initial constraint that kept the gas confined to the smaller volume.

Our task is to relate the initial and final states of the gas. The first job is to determine what function of state, if any, is the same in the two states. As far as we in the surroundings are concerned, the volume $V_2 = V_1 + V_{\text{vac}}$ is unchanged during the experiment, and the volume is thermally insulated. The removal of the constraint, in this case turning the stopcock, could, in principle, be done with a negligible amount of work. Thus the entire process involves neither work nor heat, just the removal of a constraint. We would expect the entropy to increase, since the final state is less ordered than the initial, but since $Q = 0$ and $W = 0$, we can assert that U is unchanged during this experiment. Thus *the Joule expansion of a gas into a vacuum is done at constant internal energy.*

In order to relate the initial and final temperatures and volumes for a Joule expansion, one simply puts $\Delta U = 0$ in Eq. 23-7:

$$\int_{T_1, V_1}^{T_2, V_1} C_V \, dT = -\int_{T_2, V_1}^{T_2, V_2} \left[T\left(\frac{\partial p}{\partial T}\right)_V - p \right] dV \qquad (\Delta U = 0). \quad (23\text{-}8)$$

Of course, other paths between initial and final states could have been chosen in order to fit the available heat capacity and equation of state data.

One useful way to display U-T-V data is to plot lines of constant internal energy on a T versus V diagram, as shown in Fig. 23-2. Not only can one

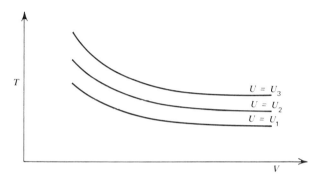

Fig. 23-2. Plot of constant energy lines for a typical gas.

find the energy by knowing T and V from such a plot but also the temperature change for various Joule expansions can be found easily. The Joule expansion from V_1 to V_2 takes the system along a line of constant U to the right, thus lowering the temperature by the amount shown.

As V increases, the lines in Fig. 23-2 level off, since dilute gases have $U = U(T \text{ only})$. Thus "ideal gases" show no temperature change on undergoing a Joule expansion, and real gases approximate this behavior at pressures up to p_c, if the temperature is at least $2T_c$.

The molecular basis for the decrease in temperature when the gas expands at constant internal energy is in the attractive well of the intermolecular potential, Fig. 18-3. When the molecules are squeezed together in the initial smaller volume, many of them are within the range of these attractive forces. When they are called upon to fill the larger final volume, fewer molecules on average can be so close together. The energy needed to pull these molecules away from one another must come from their random thermal energy, so the temperature of the gas is decreased. Clearly, at densities so small that only a negligible number of molecules are interacting with neighbors, there will be no temperature change during a Joule expansion.

The lines of constant U could be drawn from their slopes, $(\partial T/\partial V)_U$, which sometimes are given the symbol μ_J, which is called the **Joule coefficient**:

$$\mu_J \equiv \left(\frac{\partial T}{\partial V}\right)_U = -\frac{1}{C_V}\left[T\left(\frac{\partial p}{\partial T}\right)_V - p\right]. \qquad (23\text{-}9)$$

This expression follows immediately from Eq. 23-6.   However, one cannot simply integrate $\mu_J\, dV$ between V_1 and V_2 to find the temperature change in a Joule expansion.   This is because the integral would have to be performed at constant U; thus μ_J would have to be known as a function of U and V, but the right-hand side of Eq. 23-9 is likely to be known only as a function of T and V.   An analogous situation was found for isentropic processes in Prob. 22-9.   The temperature change in a constant-energy process must therefore be found from something like Eq. 23-8.

A similar situation arises in finding the entropy increase during a Joule expansion:

$$S = S(U, V) \qquad (n_i \text{ constant}), \qquad (23\text{-}10)$$

$$dS = \left(\frac{\partial S}{\partial U}\right)_V dU + \left(\frac{\partial S}{\partial V}\right)_U dV \qquad (n_i \text{ constant}). \qquad (23\text{-}11)$$

The derivatives may be identified by dividing the Gibbs equation for dU through by T:

$$dU = T\, dS - p\, dV \qquad (n_i \text{ constant}), \qquad (23\text{-}12)$$

$$dS = \frac{1}{T}\, dU + \frac{p}{T}\, dV \qquad (n_i \text{ constant}), \qquad (23\text{-}13)$$

$$S_2 - S_1 = \int_{U,\, V_1}^{U,\, V_2} \frac{p}{T}\, dV \qquad (\Delta U = 0). \qquad (23\text{-}14)$$

However, this approach requires knowing p/T as a function of the variables U and V, so that U may be held constant in the integration.   Again, the difficulty of using energy as an independent variable restricts the utility of the approach.

In order actually to find the entropy change in a Joule expansion, probably the easiest way is to find the relation between initial and final temperatures and volumes by using Eq. 23-8.   Then the entropy change may be found from Eq. 21-7 or 21-8.

One thing does emerge from Eq. 23-14, however, and that is the fact that the entropy always increases during a Joule expansion.   Since p and T are positive, the integrand of Eq. 23-14 is positive, so with $V_2 > V_1$ it follows that $S_2 > S_1$.   This is what one would expect from Clausius' inequality for this irreversible adiabatic process:

$$dS \geqslant \frac{dQ}{T} = 0 \qquad \text{(adiabatic process)}, \qquad (23\text{-}15)$$

that is, the entropy must increase as the relatively ordered state on the left of Fig. 23-1 is turned into the less ordered state on the right.

We can summarize this section with an equation analogous to Eq. 21-30, giving U as a function of T and V for fixed composition. The zero of energy is arbitrarily chosen for convenience, and different choices are made depending on the application. The energy at some low experimental temperature T_{exp} may be found relative to $U(0°K)$ from the Debye T^3 law by a procèdure analogous to that of Eqs. 21-26 to 21-29:

$$C = n\alpha T^3 \qquad (23\text{-}16)$$

$$U(T_{exp}) = U(0°K) + \int_{0°}^{T_{exp}} n\alpha T^3 \, dT, \qquad (23\text{-}17)$$

$$= U(0°K) + \tfrac{1}{4}n\alpha T_{exp}^4, \qquad (23\text{-}18)$$

$$U(T_{exp}) = U(0°K) + \tfrac{1}{4}T_{exp} C(T_{exp}). \qquad (23\text{-}19)$$

Therefore $U(T, V)$ may be written in the general form,

$$U(T, V) = U(0°K) + \tfrac{1}{4}T_{exp} C(T_{exp}) + \int_{T_{exp}, V_0}^{T, V_0} C_V(T, V_0) \, dT$$

$$+ \int_{V_0, T}^{V, T} \left[T\left(\frac{\partial p}{\partial T}\right)_V - p \right] dV + \sum_{\substack{\text{phase} \\ \text{trans}}} \Delta U_{\text{trans}} \qquad (n_i \text{ constant})$$

$$(23\text{-}20)$$

PROBLEMS

23-1. One hundred grams of benzene, C_6H_6, is vaporized at its boiling point of 80.2°C at 760 mm. The heat of vaporization is 94.4 cal/g. Calculate W, Q, and ΔU. Treat the vapor as ideal.

23-2. At a constant temperature of 25°C, 6.0 g of CO_2 gas (ideal) is expanded reversibly from 10 to 70 l. Calculate W, Q, and ΔU.

23-3. Consider the experiment shown in Fig. 23-1. Each bottle has 2.00-l capacity. The gas is oxygen, treated as ideal; the temperature is 25°C. If the initial pressure in the single bottle is 760 mm, find ΔS for the expansion.

23-4. For n moles of a gas obeying the virial equation of state $pV = nRT \times [1 + (b - a/RT)(n/V)]$, find the relation between T_1, V_1, T_2, V_2 for a Joule expansion. Assume that the heat capacity is independent of temperature.

23-5. What function of T and V is the Joule coefficient for the gas of Prob. 23-4?

23-6. Find the entropy change for the process treated in Prob. 23-4.

23-7. At 100°C, 100 l of H_2O vapor is compressed isothermally and reversibly from $\frac{1}{2}$ atm to 1 atm, and then further until the volume is 10 l. Calculate W and Q. The heat of vaporization of water is 540 cal/g.

23-8. An evacuated vessel develops a leak and fills quickly (but not in a rush) with air until the pressure inside is 1 atm. What is the temperature of the air inside before there has been time for it to equilibrate with the walls of the vessel? The air outside is at T, and C_p/C_V for the air is γ.

24 ENTHALPY CHANGES AND JOULE-THOMSON EXPANSIONS

The enthalpy is of greater practical utility than the energy because so many processes have the pressure automatically controlled by the atmosphere. This section considers the problem of relating initial and final values of the enthalpy to the initial and final temperatures and pressures for material of fixed composition. Application of the usual procedure yields

$$H = H(T, p) \qquad (n_i \text{ constant}), \tag{24-1}$$

$$dH = \left(\frac{\partial H}{\partial T}\right)_p dT + \left(\frac{\partial H}{\partial p}\right)_T dp \qquad (n_i \text{ constant}). \tag{24-2}$$

The temperature derivative is just C_p:

$$\left(\frac{\partial H}{\partial T}\right)_p = C_p, \tag{24-3}$$

and the pressure derivative is found from the Gibbs equation for dH:

$$dH = T\,dS + V\,dp. \tag{24-4}$$

Therefore,

$$\left(\frac{\partial H}{\partial p}\right)_T = T\left(\frac{\partial S}{\partial p}\right)_T + V = -T\left(\frac{\partial V}{\partial T}\right)_p + V, \tag{24-5}$$

where the last step followed from the Maxwell relation on the Gibbs free energy. If Eqs. 24-3 and 24-5 are used in Eq. 24-2, the result is

$$\boxed{dH = C_p\,dT + \left[V - T\left(\frac{\partial V}{\partial T}\right)_p\right]dp.} \tag{24-6}$$

The similarity between this result and Eq. 23-6 is noteworthy.

To find the relationship among $H_1, T_1, p_1, H_2, p_2, T_2$ for an arbitrary process (in which, of course, no phase change occurs), Eq. 24-6 may be integrated along any convenient path; for example, for the path $T_1, p_1 \rightarrow T_2, p_1 \rightarrow T_2, p_2$, the result is

$$H_2 - H_1 = \int_{T_1, p_1}^{T_2, p_1} C_p \, dT + \int_{T_2, p_1}^{T_2, p_2} \left[V - T \left(\frac{\partial V}{\partial T} \right)_p \right] dp. \qquad (24\text{-}7)$$

The ability to compute enthalpy differences is especially useful to those interested in chemical reactions, as will be seen in Sec. 30. We can write $H(T, p)$ in the general form, completely analogous to Eq. 23-20, as follows:

$$H(T, p) = H(0°\text{K}) + \tfrac{1}{4} T_{\text{exp}} \, C(T_{\text{exp}}) + \int_{T_{\text{exp}}, p_0}^{T, p_0} C_p(T, p_0) \, dT$$

$$+ \int_{p_0, T}^{p, T} \left[V - T \left(\frac{\partial V}{\partial T} \right)_p \right] dp + \sum_{\substack{\text{phase} \\ \text{trans}}} \Delta H_{\text{trans}} \qquad (n_i \text{ constant}). \quad (24\text{-}8)$$

These results permit an analysis of a phenomenon which occurs whenever fluids are forced through pipes or tubes, and which lies at the heart of the technique for liquefying gases and achieving very low temperatures, namely, the so-called **Joule-Thomson** (or Joule-Kelvin) **experiment** or effect. This is the irreversible, steady-state flow of fluid through a throttle, a constriction, or some kind of orifice. An example is shown in Fig. 24-1. There is a

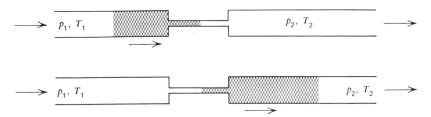

Fig. 24-1. A given amount of material undergoing a Joule-Thompson expansion.

steady flow of fluid, so the two sides of the constriction remain filled with fluid, and the state of the fluid to the left of the constriction remains constant in time, as does that to the right. The walls of the vessel are presumed insulated in the analysis, and no heat is presumed to flow in the direction of fluid motion (horizontally in Fig. 24-1, as drawn).

We want to relate the initial and final states of the fluid. As always, the first job is to determine what function of state, if any, is the same in the two states. In order to do this there is a slight question of what to call the system. If it is just the contents of some fixed part of the container, then since a steady state is being maintained, nothing is changing and there is no interest in the problem.

With a flowing system, the obvious question is what happens to a given amount of material as it flows through. So, as shown in Fig. 24-1, we focus on a given amount of material as it is pushed through the throttle. The energy change for this material is given by the first law:

$$E_2 - E_1 = Q + W. \tag{24-9}$$

The presumptions made about heat flow imply that Q is zero for the fluid as it undergoes the expansion. The work is easily calculated. On the left of the orifice, the surroundings do work on the fluid:

$$W_{\text{left}} = -\int p(V)\, dV = -\int p_1\, dV = -p_1 \int dV = -p_1(0 - V_1) = p_1 V_1. \tag{24-10}$$

Since the pressure is constant at p_1, the work is simply the negative of the pressure times the volume change. On the right, work is done on the surroundings by the fluid:

$$W_{\text{right}} = -\int p(V)\, dV = -\int p_2\, dV = -p_2 \int dV = -p_2(V_2 - 0) = -p_2 V_2. \tag{24-11}$$

The sum of Eqs. 24-10 and 24-11 goes into Eq. 24-9 as W, and $Q = 0$:

$$E_2 - E_1 = p_1 V_1 - p_2 V_2, \tag{24-12}$$

$$U_2 + p_2 V_2 = U_1 + p_1 V_1, \tag{24-13}$$

$$H_2 = H_1. \tag{24-14}$$

In Eq. 24-13, the total energy was replaced by the internal energy. This is an approximation which is reasonable for fluids not flowing very fast, but since there is kinetic energy of motion for this flowing fluid, it is not quite correct. Thus *the Joule-Thomson expansion occurs at almost constant enthalpy; the process is nearly isenthalpic.*

One can now go back to Eq. 24-7 and put $\Delta H = 0$ to get the relationship between the initial and final temperatures and pressures for a fluid undergoing a Joule-Thomson expansion:

$$\int_{T_1,\, p_1}^{T_2,\, p_1} C_p\, dT = \int_{T_2,\, p_1}^{T_2,\, p_2} \left[T\left(\frac{\partial V}{\partial T}\right)_p - V \right] dp \qquad (\Delta H = 0). \tag{24-15}$$

Other paths of integration of Eq. 24-6 could just as well have been chosen, if convenient.

It is helpful to display H-T-p data either as isenthalps on a plot of T against p or of isotherms on an H-p plot. A typical example of the former is shown in Fig. 24-2. The slope of an isenthalp is called the *Joule-Thomson coefficient* μ_{JT}:

$$\left(\frac{\partial T}{\partial p}\right)_H \equiv \mu_{JT} = \frac{1}{C_p}\left[T\left(\frac{\partial V}{\partial T}\right)_p - V\right]. \tag{24-16}$$

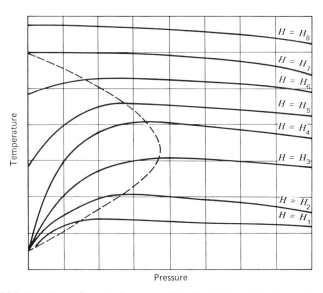

Fig. 24-2. Isenthalps for a typical gas. The dashed line is the inversion curve.

Upon expansion through a throttle or porous plug, the state of the gas moves to the left along a line of constant H from the higher to lower pressure. For sufficiently high pressures, the result is a warming of the gas. However, for low enough pressures the gas is cooled during the expansion. The temperature at which μ_{JT} changes from positive to negative is called the *inversion temperature*. At the inversion temperature, $\mu_{JT} = 0$, or

$$T\left(\frac{\partial V}{\partial T}\right)_p = V \quad \text{(inversion temperature)}, \tag{24-17}$$

which gives a relationship between T and p which is shown by the dashed line of Fig. 24-2. In order to get an idea of the scale of Fig. 24-2, we note that the maximum inversion temperature of nitrogen is about 620°K, of hydrogen about 200°K, and of helium about 35°K. Pressures at these maxima are of the order of 400, 160, and 40 atm respectively.

Perhaps the most important application of the Joule-Thomson effect is in liquefying gases. For example, hydrogen which boils at about 20°K, and even helium which boils at 4.2°K, can quite readily be liquefied. It is clear from Fig. 24-2 that, if a gas is expanded through a throttle, it might very well end up at a lower temperature. If it entered the expansion vessel at or below its inversion temperature, cooling would be certain. Of course, the temperature does not fall directly to anything like $4.2°K°$! Instead, what is done is to create a steady-state countercurrent heat exchanger in which the expanded cooler gas is used to cool the incoming compressed gas. Then when that cooled gas is expanded, the resulting temperature is lower still. This process is continued until finally gas at room temperature (or perhaps at the temperature of liquid nitrogen if something like hydrogen or helium is being liquefied) is introduced into the liquefier at high pressure. Out come two products: a small fraction z of the input gas is liquefied and emerges at 1 atm of pressure. The rest emerges as gas at 1 atm of pressure. A sketch of the mechanism by which this is achieved is given in Fig. 24-3.

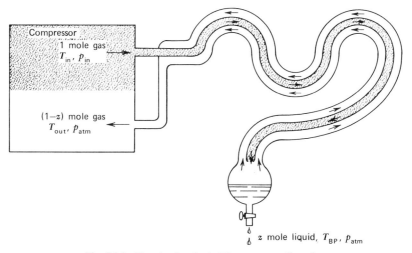

Fig. 24-3. Sketch of a Joule-Thomson gas liquefier.

The nozzle, at which the bulk of the pressure drop occurs, is at the collecting basin for the liquefied gas. The liquid is drawn off into Dewar flasks at the same rate it is formed. In the long heat exchanger incoming compressed gas is cooled by outgoing expanded gas. At the compressor the expanded gas is recompressed to the pressure p_{in}. Of course, in the process of compression it heats up, so it then is cooled back to the temperature T_{in} before being recirculated. The entire right-hand side of the unit as pictured in Fig. 24-3 is carefully insulated thermally.

The process of taking a mole of gas through the right-hand side of Fig. 24-3 is isenthalpic, and there is nothing in our proof of Eq. 24-14 which breaks down even if some of the fluid changes phase in the course of the expansion. Thus it is easy to compute the maximum value of z. If we know the molar enthalpy of the gas at T_{in}, p_{in} and at T_{out} and p_{atm}, and also that of the liquid at its boiling point T_{BP} and p_{atm}, then an enthalpy balance (enthalpy in = total enthalpy out) yields the maximum value of z, which would be achieved if there were no heat leakage:

$$1 h_{gas}(T_{in}, p_{in}) = (1 - z) h_{gas}(T_{out}, p_{atm}) + z h_{liq}(T_{BP}, p_{atm}), \qquad (24\text{-}18)$$

$$z_{max} = \frac{h_{gas}(T_{out}, p_{atm}) - h_{gas}(T_{in}, p_{in})}{h_{gas}(T_{out}, p_{atm}) - h_{liq}(T_{BP}, p_{atm})} \qquad (24\text{-}19)$$

Another important use of the Joule-Thomson experiment is to use experimental measurements of C_p and μ_{JT} to find the difference between $T(\partial V/\partial T)_p$ and V. For an ideal gas this would be zero, so its magnitude permits a study of the effects of nonideality in real gases. This is of special help in correcting real gas thermometers for the effects of the nonideality of the gas under the finite pressure necessary for use.

PROBLEMS

24-1. Prove there is no temperature change for an ideal gas undergoing a Joule-Thomson expansion.

24-2. One mole of ammonia (considered to be an ideal gas) initially at 25°C and 1 atm of pressure is heated reversibly at constant pressure until the volume has trebled. Calculate (a) Q; (b) W; (c) ΔH; (d) ΔU; (e) ΔS. The molar heat capacity of NH_3 at constant pressure is $6.189 + 0.007887T - 7.28 \times 10^{-7}T^2$ cal/deg-mole, where the temperatures are in °K.

24-3. What is the final temperature of the system obtained from mixing 100 g of water at 50°C with 10 g of ice at 0°C? The heat of fusion of ice is 80 cal/g.

24-4. Calculate ΔS for the Joule-Thomson expansion of 2.000 g of air (average molecular weight = 29) from 3.0 to 1.0 atm. Treat the air as an ideal gas. The initial temperature is 25°C.

24-5. The power of the sun's radiation incident on the earth is approximately 1.2 kw/m² on a clear day. If the temperature of a lake stays constant and all the cooling is by evaporation, how much is the water level lowered in 12 hr? The heat of vaporization is 540 cal/g.

24-6. For nitromethane at 15°K, the molar heat capacity is 0.89 cal/deg-mole. Assume the Debye T^3 law holds for temperatures below 15°K, and calculate the molar entropy at 15°K and the molar enthalpy relative to its value at 0°K.

24-7. At 0°C ice absorbs 80 cal/g in melting, water absorbs 597 cal/g in vaporizing. What is the heat of sublimation of ice at 0°C?

24-8. A substance of molecular weight 42 has heat of sublimation at −30°C and 1 atm of 7300 cal/mole. What is ΔH and ΔU for the condensation of $\frac{1}{2}$ g of the substance from vapor to solid at −30°C and 1 atm?

24-9. A process taking a system from 2 atm and 8 l to $\frac{1}{2}$ atm and 12 l has $\Delta U = 20$ l-atm. What is ΔH?

24-10. If a gas with $C_p = \frac{7}{2}nR$ undergoes a Joule-Kelvin expansion from 100 to 1 atm, starting at 25°C, what is the final temperature? The equation of state is $p(V - nb) = nRT$, where b is 0.045 l/mole.

24-11. For n moles of a gas obeying the virial equation of state $pV = nRT + n(b - a/RT)p$, find the relationship between initial and final temperatures and pressures for a Joule-Kelvin expansion.

24-12. The gas described in Prob. 24-11 has a unique inversion temperature. What is it?

24-13. Show that in a small Joule-Thomson expansion of a fluid the temperature change is $1 - 1/T\alpha$ times that obtained from the corresponding isentropic pressure drop.

24-14. Liquid benzene is expanded adiabatically through a porous plug. What is ΔT/atm change for benzene at $300°K$? The molar volume is 88.8 ml, $\alpha = 1.24 \times 10^{-3}$ deg^{-1}, and $C_p = 33.2$ cal/deg-mole.

24-15. Calendar's equation of state for 1 mole of steam is $p(v - b) = RT - Ap/T^{3.33}$, where $A = 2.00 \times 10^8$ l-deg$^{3.33}$/mole and $b = 0.018$ l/mole. What is the inversion temperature of steam?

24-16. On a plot of $B(T)$ versus T, Fig. 18-5, indicate the inversion temperature and the so-called **Boyle temperature** (the temperature at which Boyle's law, $pV = $ constant, is obeyed) if one can neglect contributions from higher virial coefficients.

24-17. Prove from Eq. 24-19 that the greatest fraction of entering gas will be liquefied if the pressure of the entering gas is adjusted so that p_{in}, T_{in} is a point on the inversion curve.

24-18. In an efficient Joule-Kelvin gas liquefier, O_2 enters at $298°K$ and 200 atm, gas emeges at $298°K$ and 1 atm, and liquid oxygen emerges at its boiling point of $90°K$. What fraction of the entering gas is liquefied? For oxygen, $C_p = 6.95$ cal/deg-mole; ΔH of vaporization is 1629 cal/mole at $90°K$, and the throttling experiment on a mass of gas from 200 to 1 atm reduces its temperature from 298 to about $248°K$.

25 FREE ENERGY CHANGES AND PHASE EQUILIBRIA

FREE ENERGY CHANGES

In addition to applications of the type we encounter later, calculations of free energy changes are often requested on thermodynamics examinations. Values of ΔA and ΔG are usually obtained either from integrating the Gibbs equations

$$dA = - S \, dT - p \, dV \qquad (n_i \text{ constant}), \qquad (25\text{-}1)$$

$$dG = - S \, dT + V \, dp \qquad (n_i \text{ constant}), \qquad (25\text{-}2)$$

or from the definitions

$$\Delta A = \Delta U - \Delta(TS), \tag{25-3}$$

$$\Delta G = \Delta U - \Delta(TS) + \Delta(pV) = \Delta H + \Delta(pV). \tag{25-4}$$

The $\Delta(TS)$ notation means the change $T_2 S_2 - T_1 S_1$ in the TS product.   It is *not*, as students sometimes write, $T\,\Delta S + S\,\Delta T$, a form imprinted mentally from its use with infinitesimals, but meaningless for finite changes (unless, of course, either ΔS or ΔT is zero).

INTRODUCTION TO PHASE EQUILIBRIA

Now we return to the plot of p versus v and T, Fig. 18-1, and focus on the equilibria between different phases for single-component systems.   For this purpose it is convenient to look at the two-dimensional projection of Fig. 18-1 similar to Fig. 18-2, only projected onto the p-T plane.   This is shown in Fig. 25-1, and it is visualized more clearly if it is compared with Figs. 18-1 and 18-2.   A pair of values p and T designates a point in Fig. 25-1; the region in which the point lies establishes the equilibrium phase of the material.   The lines of Fig. 25-1 represent the locus of conditions under which the two phases, solid-vapor, solid-liquid, or liquid-vapor, can coexist.

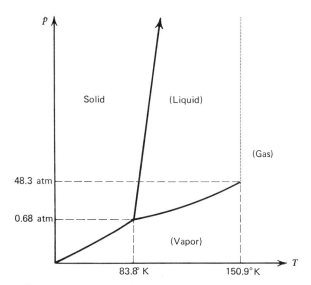

Fig. 25-1. Two-dimensional projection of Figure 18-1 onto the p-T plane.   The values given are for argon, showing the extent to which the scale of the diagram has been distorted for the sake of clarity.

There is a unique pressure and temperature at which all three phases can coexist—the triple point for the substance.   The vapor-liquid line stops at the critical point.   As was discussed in Sec. 18, the definitions of liquid and gas are arbitrary except on a phase-coexistence line.   The whole liquid-gas-vapor region could be called simply the fluid state.   Evidence indicates that the solid-liquid line continues to enormously high pressures without stopping at a solid-liquid critical point.   However, if the pressure gets so high as actually to destroy the structures of the atoms and molecules making up the substance, then the distinction between solid and liquid most likely disappears.   The experimental evidence is unclear and hard to obtain.

The pressure of the vapor that would be in equilibrium with the pure liquid or solid at a particular temperature is called the **vapor pressure** of the liquid or solid (sometimes called the **sublimation pressure** for solids).   Thus the solid-vapor and liquid-vapor lines are just plots of vapor pressure as a function of temperature.

The concept of vapor pressure deserves elaboration.   Suppose some of the pure solid or liquid is sealed into part of a vessel which is otherwise evacuated, as shown in Fig. 25-2a.   The vessel is placed in a thermostat, and

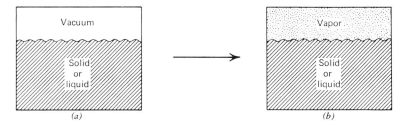

Fig. 25-2. (a) Pure solid or liquid sealed in evacuated vessel at fixed temperature. (b) Pressure of vapor increases until equilibrium vapor pressure is reached.

the only thing on which the final equilibrium state can depend is the temperature.   The equilibrium pressure finally reached in the vessel is the vapor pressure of the pure solid or liquid at that temperature.   Once it has been measured by something resembling Fig. 25-2b, the vapor pressure can be considered to be one of the properties (a function of T) of a solid or liquid, regardless of whether any vapor happens to be present.

On the molecular scale, what happens in the process of Fig. 25-2 is easy to visualize.   Molecules near the surface of the condensed phase in Fig. 25-1a are held in by the intermolecular forces of attraction.   However, because of the random thermal motion of the particles, occasionally one molecule will be given an extra wallop of energy due to fortuitous collisions with neighbors, and it will be able to break loose from the condensed mass and fly freely into the vapor. As the concentration in the vapor builds up, the rate of collisions of molecules from the vapor with the condensed material increases. Many of these are sticking collisions, in that the molecule from the gas rejoins the condensed phase and has lost the energy needed to bounce back into the vapor. Thus there is migration of particles in both directions, from the condensed phase to the vapor and back again. Finally, a dynamic equilibrium is reached in which the vapor becomes dense enough for the rates of each process to be the same. The probability of enough thermal energy being given in a fluctuation to a particle near the surface to permit it to escape increases rapidly as the temperature is increased. This effect more than counterbalances the increased rate of collisions of molecules in the vapor with the surface; thus the vapor pressure increases with temperature.

CHEMICAL POTENTIAL SURFACES

We would like to be able to express the phase-coexistence lines as functions of thermodynamic properties of the substance and thus to understand the phase diagram. At any given T and p, the material *could* exist in solid, liquid, or vapor phase; yet it *does* exist in just one. It must be the one with the lowest value for the chemical potential at that T and p, since matter flows (by melting, freezing, condensing, subliming, etc.) from states of high to low chemical potential. Thus the stable state at a particular T and p is that with the lowest value of $\mu(T, p)$ for the material. Where two phases can coexist in equilibrium, the chemical potentials of the material in each phase at that T and p must be equal; then, of course, μ of the other phase must lie above the μ for the two coexisting phases. The particular T and p where the values of μ happen to be the same for all three phases is the triple point.

Therefore, if one can express μ as a function of T and p for each phase— solid, liquid, and vapor—one may simply plot these three μ surfaces as functions of T and p. Then, for a given value of T and p, whichever surface lies lowest represents the stable state of the material. Along the line where the lowest two μ surfaces intersect, those two phases may coexist at equilibrium.

It was shown in Sec. 13 that for a single-component system μ is just the molar Gibbs free energy: $\mu = G/n$. For a single-component system with $n = 1$, the Gibbs equation, Eq. 13-23, for the Gibbs free energy is

$$\boxed{d\mu = -s\,dT + v\,dp}$$ (single component).   (25-5)

This is just an expression of the Gibbs-Duhem equation, Eq. 13-32, for 1 mole of a single-component substance.

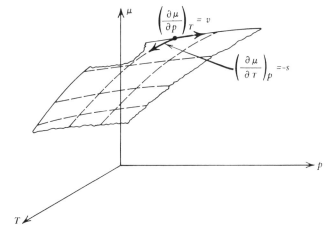

Fig. 25-3. Typical plot of μ against T and p.

In Fig. 25-3 a typical bit of the μ surface for a phase is pictured. The principal slopes of this surface are identified from Eq. 25-5 as

$$\left(\frac{\partial \mu}{\partial T}\right)_p = -s, \qquad \left(\frac{\partial \mu}{\partial p}\right)_T = v. \tag{25-6}$$

We note that $-s$ is always negative and gets more so as T goes up and less so as p goes up. The v is positive, small and nearly constant for solids and liquids, relatively large for gases and increasing as T goes up and decreasing as p goes up.

If we take a simple substance with three possible phases and construct μ surfaces for each phase, the lowest lying surface represents the stable phase. It is easy to see why the vapor phase, if it exists at all, must exist at the bottom (lowest pressures) in Fig. 25-1. Suppose one slices through the three μ surfaces parallel to the μ-T plane. Also, suppose $v_{liq} > v_{sol}$, which is the case for most substances except water, bismuth, and antimony. If the isotherm is below the temperature of the triple point, the μ-p plot looks something like Fig. 25-4a. If the isotherm is between the triple point and the critical point, the plot of μ versus p resembles Fig. 25-4b. Since v_{vap} is greater than either v_{liq} or v_{sol}, the vapor line is steeper sloping than either the liquid or solid line. Thus if it lies below these lines, it must do so at low pressures, since its greater slope inevitably carries it above them at sufficiently high pressures. Similarly, since $v_{liq} > v_{sol}$, if the liquid phase is to exist at all, it must be at lower pressures than the solid. The solid, once it becomes stable, remains so as the pressure increases, because its molar volume is the least. Thus the solid-liquid line in Fig. 25-1 slopes steeply upward from the triple point rather than steeply down to it, so that moving straight up from

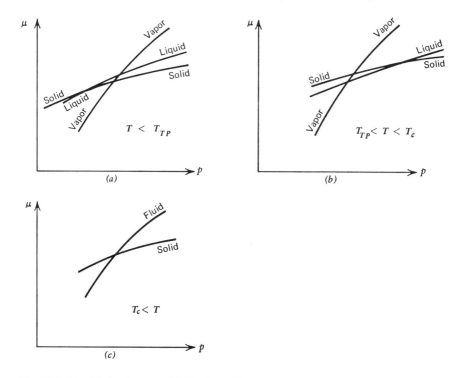

Fig. 25-4. Possible isotherms.   (a) $T < T_{TP}$; (b) $T_{TP} < T < T_c$; (c) $T > T_c$.   Slope of each line is molar volume of that phase at the particular T and p.

some point to the right of the triple point takes one from vapor to liquid to solid in that order.   For temperatures above T_c, the liquid and gas μ surfaces merge to become a single fluid surface, as shown in Fig. 25-4c.

For water, bismuth, and antimony, $v_{sol} > v_{liq}$, and the solid-liquid line slopes steeply down to the triple point, as shown in Fig. 25-5.   This remarkable property of water, that is, in having its liquid volume less than its solid volume, is responsible for the fact that ice floats.   Thus ponds do not freeze solid in the winter and life can survive in the liquid at the bottoms.   No doubt, this fact has permitted some forms of life to arise and evolve which would never have been possible were $v_{sol} < v_{liq}$ for water.

An analysis of the result of creating constant-pressure surfaces by slicing through the μ surfaces parallel to the μ-p plane is similar to the foregoing. The slopes of the three resulting lines are given by $(\partial \mu / \partial T)_p = -s$, with plots resembling those of Fig. 25-6.   Since the entropies of the phases increase as one goes from solid to liquid to gas, vapor must be the stable phase at high temperatures and solid at low.   If the liquid is stable at all at

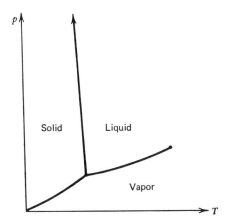

Fig. 25-5. Typical phase diagram for water, bismuth, or antimony.

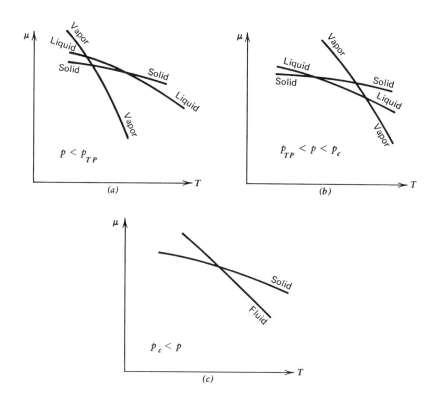

Fig. 25-6. Possible isobars.   (a) $p < p_{TP}$; (b)$p_{TP} < p < p_c$; (c) $p > p_c$.   Slope of each line is negative molar entropy of that phase at the particular T and p.

the pressure chosen, its range of stable temperatures must lie between those of the solid and the vapor.

The kinds of phase transitions we have been discussing involve discontinuous changes in the molar entropy and the molar volume during the course of the phase transition. Since $-s$ and v are the *first* derivatives of the chemical potential, these phase changes are called **first-order phase transitions**. Second-order phase transitions are also known; with them the discontinuous changes are not in entropy or volume; these functions remain continuous across the transition. Instead, there are discontinuous changes in properties such as heat capacities and compressibilities, which are *second* derivatives of the chemical potential.

CLAPEYRON AND CLAUSIUS-CLAPEYRON EQUATIONS

The phase-coexistence lines of Fig. 25-1 or 25-5 may be related mathematically to the thermodynamic properties of the phases. This permits quantitative predictions of great utility, as the problems of this section will suggest. The phase-coexistence lines (or phase-transition lines) are characterized by the fact that along them the chemical potentials of the substance in the two phases are equal:

$$\mu_1(T, p) = \mu_2(T, p) \qquad \text{(phase-coexistence line)}. \qquad (25\text{-}7)$$

Thus, as T and p change along a coexistence line,

$$d\mu_1 = d\mu_2 \qquad \text{(phase-coexistence line)}, \qquad (25\text{-}8)$$

$$-s_1 \, dT + v_1 \, dp = -s_2 \, dT + v_2 \, dp \qquad \text{(phase-coexistence line)}, \quad (25\text{-}9)$$

where Eq. 25-5 was used. If we solve Eq. 25-9 for the ratio between dp and dT along the phase-coexistence line, we obtain

$$\left. \frac{dp}{dT} \right]_{\substack{\text{along a} \\ \text{phase trans} \\ \text{line}}} = \frac{s_2 - s_1}{v_2 - v_1} = \frac{\Delta s}{\Delta v} = \frac{\Delta S}{\Delta V} = \frac{\Delta H_{\text{trans}}}{T \, \Delta V_{\text{trans}}}, \qquad (25\text{-}10)$$

where substitution of $\Delta H/T$ for ΔS was made from Eq. 21-20. This is the so-called **Clapeyron equation** for the slope of a phase-transition or phase-coexistence line. It correctly gives the slope of any of the three lines of Fig. 25-1 or 25-5. It is used in the form of Eq. 25-10, mainly for solid-liquid equilibria (or solid-solid equilibria for those substances having several different solid phases), because the Clausius-Clapeyron equation, developed in the following, or some modification thereof is more convenient for equilibria where one of the phases is gas.   The best way to remember any of these equations is to derive them.

If one of the coexisting phases is the vapor, then (unless one is in the immediate vicinity of the critical point or is at extraordinarily high pressures) the volume of the condensed phase is negligible compared to that of the vapor (for example, the molar volume of liquid water is only about 0.018 l):

$$\frac{dp}{dT}\bigg]_{\text{trans}} = \frac{\Delta H}{TV_{\text{vap}}} \qquad (V_{\text{vap}} \gg V_{\text{sol}} \text{ or } V_{\text{liq}}). \qquad (25\text{-}11)$$

In order to integrate this equation, the p and T dependence of ΔH and V_{vap} are needed.   One can use either tables of volumetric data or an appropriate equation of state for $V(T, p)$.   The ideal-gas law is often used, at least as a reasonable first approximation up to $p = \frac{1}{2}p_c$.   If $V = nRT/p$ is inserted into Eq. 25-11, it becomes

$$\frac{dp}{dT}\bigg]_{\text{trans}} = \frac{p\,\Delta H}{nRT^2} \qquad (V_{\text{vap}} \gg V_{\text{sol}} \text{ or } V_{\text{liq}} \text{ ; ideal-gas law}), \qquad (25\text{-}12)$$

The variables can be separated,

$$\frac{dp}{p} = \frac{\Delta H\,dT}{nRT^2} \qquad (V_{\text{vap}} \gg V_{\text{sol}} \text{ or } V_{\text{liq}}; \text{ ideal-gas law}), \qquad (25\text{-}13)$$

and if the T and p dependence of ΔH is neglected, Eq. 25-13 can be integrated to give

$$\boxed{\begin{aligned} &\ln p = -\frac{\Delta H}{nRT} + \text{constant} \\[2mm] &\text{or} \\[2mm] &\ln \frac{p_2}{p_1} = -\frac{\Delta H}{nR}\left(\frac{1}{T_2} - \frac{1}{T_1}\right) \\[2mm] &\quad = \frac{\Delta H(T_2 - T_1)}{nRT_2 T_1} \end{aligned}}$$

$$(25\text{-}14)$$

(ideal gas law;
ΔH independent of p and T;
$V_{\text{vap}} \gg V_{\text{sol}}$ or V_{liq}).

$$(25\text{-}15)$$

These equations are sometimes called the **Clausius-Clapeyron equation.** Equation 25-14 is the equation of either of the two lines involving vapor in Fig. 25-1 or 25-5, to the approximation of neglecting the volume of the condensed phase and use of the ideal-gas law.   Equation 25-15 is the relation between any pair of points on one of these two lines, to the same degree of approximation.   In the dilute-gas range where the ideal-gas law is reasonable, the fact that ΔH is somewhat dependent on temperature will show up on a plot of $\log p$ versus $1/T$ as a line which is not quite straight. The slope of that line at any point is shown by Eq. 25-14 to be $-\Delta H(T, p)/2.303nR$.

Example 25-1. Early drafts of this book stated that "ice skating is made possible because just under the narrow blade of the skate, the pressure on the ice is suddenly increased. This lowers the melting point just enough for a thin film of ice to melt, and the liquid water lubricates the passage of the skate." Check the soundness of that statement for an ice skate blade 2 mm wide with 10 cm of blade in contact with the ice by estimating the lowest temperature at which a 150-lb man could skate on both skates at once. The density of ice is 0.9168, the heat of fusion of ice is 79.7 cal/g; 1 lb = 2.2 kg.

Solution: For small temperature changes we can use the Clapeyron equation, Eq. 25-10, in the form

$$\Delta T = \frac{T \, \Delta V \, \Delta p}{\Delta H}. \tag{25-16}$$

The pressure change is the gravitational force exerted on the man divided by the area of the two skate blades:

$$\Delta p = \frac{(150 \text{ lb})(\text{kg})(1000 \text{ g})(980 \text{ cm})}{(2.2 \text{ lb}) (\text{kg}) \quad (\text{sec}^2)(2)(10 \text{ cm})(0.2 \text{ cm})}. \tag{25-17}$$

The volume change for the melting of 1 g of ice is

$$\Delta V = \frac{(1 \text{ g})(\text{ml})}{(1 \text{ g})} - \frac{(1 \text{ g})(\text{ml})}{(0.9168 \text{ g})} = -\frac{0.0832 \text{ ml}}{0.9168 \text{ g}}. \tag{25-18}$$

Thus Eq. 25-16 is

$$\Delta T =$$

$$\frac{-(273 \text{ deg})(0.0832 \text{ ml})(150)(1000)(980)(\text{dyne})}{(0.9168) \quad (2.2)(2)(10)(0.2) \, (\text{cm}^2)(79.7 \text{ cal})(4.184 \times 10^7 \text{ ergs})} \cdot \frac{1 \text{ cal}}{},$$

$$\tag{25-19}$$

$$\Delta T = -0.124°C. \tag{25-20}$$

This explains why the foregoing paragraph was removed from this book. Yet, highly reputable textbooks give this same argument; see Prob. 25-2 for a restatement which suggests a resolution of the puzzle.

Example 25-2. The vapor pressure of water at 0°C is 4.58 mm Hg. Use the Clausius-Clapeyron equation to calculate the normal boiling point of water. Treat the heat of vaporization as 595 cal/g, independent of temperature.

Solution: We know $T_1 = 273°K$, $p_1 = 4.58$ mm, $p_2 = 760$ mm, and we want T_2. From Eq. 25-15,

$$\frac{1}{T_2} = \frac{1}{T_1} - \frac{nR}{\Delta H} \ln \frac{p_2}{p_1}, \tag{25-21}$$

$$= \frac{1}{273 \text{ deg}} - \frac{(1 \text{ mole})(1.987 \text{ cal})(\text{g})(2.303)}{(\text{deg mole})(595 \text{ cal})(18 \text{ g})} \log \frac{760}{4.58} \tag{25-22}$$

$$= 0.00366 \text{ deg}^{-1} - 0.000948 \text{ deg}^{-1}, \tag{25-23}$$

$$T_2 = 369°K = 96°C. \tag{25-24}$$

This answer is within $4°$ of the correct result. What would be the sublimation pressure of ice at $0°C$?

THE KIRCHHOFF EQUATION

One often needs to know Δh for a phase transition along a whole segment of a phase-transition line. This usually must be based on only a single Δh value plus some other thermodynamic information. The temperature dependence of the enthalpy is straightforward; Equation 24-7 shows $h(T')$ to be related to its value at T_0 by

$$h(T') = h(T_0) + \int_{T_0}^{T'} c_p \, dT \qquad (p \text{ constant}). \tag{25-25}$$

If we write Eq. 25-25 for each phase involved in the transition and subtract the two equations, we get

$$\Delta h(T') = \Delta h(T_0) + \int_{T_0}^{T'} \Delta c_p \, dT \qquad (p \text{ constant}). \tag{25-26}$$

If Δc_p is essentially constant over the interval, this becomes

$$\Delta h(T') = \Delta h(T_0) + \Delta c_p(T' - T_0) \qquad (p \text{ and } \Delta c_p \text{ constant}). \tag{25-27}$$

Otherwise, temperature corrections to the heat capacities must be made. Equation 25-26 is called the Kirchhoff equation.

These equations are derived for constant pressure, and along phase-transition lines the pressure is not constant. There is only one degree of freedom, which may be T, but p changes automatically when T changes, in order that the system point stay on the transition line. Therefore one should integrate the total derivative of Δh in Eq. 25-26, rather than the partial derivative:

$$\frac{d\,\Delta h}{dT} = \left(\frac{\partial\,\Delta h}{\partial T}\right)_p + \left(\frac{\partial\,\Delta h}{\partial p}\right)_T \left(\frac{dp}{dT}\right)_{\text{trans}}, \tag{25-28}$$

$$= \Delta c_p + \left[\Delta v - T\left(\frac{\partial\,\Delta v}{\partial T}\right)_p\right]\frac{\Delta h}{T\,\Delta v}, \tag{25-29}$$

$$\frac{d\,\Delta h}{dT} = \Delta c_p + \frac{\Delta h}{T} - \Delta h\left(\frac{\partial \ln \Delta v}{\partial T}\right)_p, \tag{25-30}$$

where Eq. 24-6 was used for $(\partial h/\partial p)_T$ and the Clapeyron equation for dp/dT.

Now for sublimation and vaporization well below the critical point, Δv is nearly the molar volume of the vapor, and

$$\left(\frac{\partial \ln \Delta v}{\partial T}\right)_p = \left(\frac{\partial \ln v_v}{\partial T}\right)_p = \frac{1}{T} + \left(\frac{\partial \ln z_v}{\partial T}\right)_p \qquad (v_v \gg v_{\text{cond}}). \tag{25-31}$$

If this is used in Eq. 25-30, the result is

$$\frac{d\,\Delta h}{dT} = \Delta c_p - \Delta h\left(\frac{\partial \ln z_v}{\partial T}\right)_p \qquad \text{(sublimation or vaporization, } v_v \gg v_{\text{cond}}),$$

$$\tag{25-32}$$

and usually the last term can be neglected, thus confirming Kirchhoff's equation, Eq. 25-26, for sublimation and vaporization.

However, for fusion, where both molar volumes are small, $\partial v/\partial T$ is smaller yet, and the last term of Eq. 25-30 can be neglected:

$$\frac{d\,\Delta h}{dT} = \Delta c_p + \frac{\Delta h}{T} \qquad \text{(condensed, incompressible phases)}, \tag{25-33}$$

or

$$\Delta h(T') = \Delta h(T_0) + \int_{T_0}^{T'} \left(\Delta c_p + \frac{\Delta h}{T}\right) dT$$

$$\text{(condensed, incompressible phases)}. \tag{25-34}$$

Equation 25-34 is much more precise than Eq. 25-26 for transitions involving only condensed phases. The last term of Eq. 25-33 may represent 40% of the total contribution to the change in Δh (see Prob. 25-16).

PROBLEMS

25-1. One mole of toluene is vaporized at its boiling point, 111°C, against the standard atmospheric pressure. The heat of vaporization at this temperature is 86.5 cal/g. Calculate (a) W; (b) Q; (c) ΔH; (d) ΔU; (e) ΔG; (f) ΔS.

25-2. A reputable textbook in calculating the problem posed in Example 25-1 states that Δp is approximately 1.8×10^9 dynes/cm². If that book is correct, what is ΔT? What is implied by that value of Δp about the way ice skates are sharpened?

25-3. The heats of vaporization and of fusion of water are 595 cal/g and 79.7 cal/g at 0°C. The vapor pressure of water at 0°C is 4.58 mm. Calculate the sublimation pressure of ice at −15°C, assuming that the enthalpy changes are independent of temperature.

25-4. Ten liters of air was bubbled through carbon tetrachloride at 20°C and 1 atm of pressure. The loss in weight of the liquid was 8.698 g. Calculate the vapor pressure of CCl_4.

25-5. Propene has the following vapor pressures: at 150°K, $p = 3.82$ mm; at 200°K, $p = 198.0$ mm; at 250°K, $p = 2074$ mm; at 300°K, $p = 10,040$ mm. From these data calculate (a) the heat of vaporization and (b) the vapor pressure at 225°K by a graphical method.

25-6. Liquid mercury has a density of 13.690 g/ml, and solid mercury has a density of 14.193 g/ml, both being measured at the melting point, −38.87°C, under a pressure of 1 atm. The heat of fusion is 2.33 cal/g. Calculate the melting points of mercury under a pressure of (a) 10 atm, and (b) 3540 atm.

25-7. For diethyl ether (molecular weight = 74) at its boiling point, 34.5°C, the heat of vaporization is 88.39 cal/g. What is its vapor pressure at 100°C?

25-8. What is the boiling point of water on a mountain where the barometer reading is 600 mm? The heat of vaporization of water may be taken to be 9.72 kcal/mole.

25-9. Thirty grams of liquid water is vaporized reversibly at 100°C and 1 atm. The vapor is expanded reversibly and isothermally to $\frac{1}{3}$ atm. ΔH_{vap} for H_2O is 10.5 kcal/mole. Calculate W, Q, ΔS, ΔU, ΔH, ΔA, and ΔG.

25-10. Why are the lines in Figs. 25-4 and 25-6 not straight, and why do they all have negative curvature?

25-11. A gas obeys the equation of state: $pV = nRT + nbp$, where b is a constant. If the n moles of the gas expands from V_1 to V_2 reversibly at constant temperature T, calculate (a) ΔA; (b) ΔG; (c) ΔS; (d) ΔU; (e) W, (f) Q.

25-12. Suppose the expansion of Prob. 25-11 was not reversible, but was into a vacuum instead, that is, a Joule expansion from V_1 to V_2. How would the results of Prob. 25-11 be changed?

25-13. The vapor pressure, in torr (that is, in mm Hg), of solid ammonia is given by $\ln p = 23.03 - 3754/T$ and that of liquid ammonia by $\ln p = 19.49 - 3063/T$. (a) What is the temperature of the triple point? (b) What are the heats of sublimation and vaporization? (c) What is the heat of fusion at the triple point?

25-14. Suppose a certain alloy exists in two different solid phases, α and β. The phase transition is first order and $v_\alpha < v_\beta$; $s_\alpha > s_\beta$. Sketch a μ-p isotherm showing both the α line and the β line and how they intersect. Sketch a μ-T isobar showing both the α and β lines. Sketch a short segment of the p-T line representing $\mu_\alpha = \mu_\beta$ and state in words what its physical meaning is.

25-15. Ten grams of water at 20°C is mixed with 10 g water at 80°C in an insulated vessel. Calculate W, Q, ΔU, ΔH, ΔS, ΔA, and ΔG. The molar entropy of water, if it is desired, is 18.17 eu at 50°C.

25-16. For the process of Prob. 21-21, what is the change in Δh due to the lowered temperature?

25-17. The transition from gray to white tin has $\Delta H = 0.50$ kcal/mole and $\Delta S = 1.745$ eu/mole, presumed independent of T. At what temperature do these forms of tin coexist?

25-18. One mole of KBr at 300°C and 1 atm is compressed adiabatically and reversibly to 6000 atm. Calculate Q, W, ΔU, ΔH, ΔS, ΔA, and ΔG. KBr has

$v = 43.28$ cm³/mole at 1 atm, $\alpha = 12.00 \times 10^{-5}$ deg⁻¹, $\beta = 6.75 \times 10^{-6}$ atm⁻¹, and $C_p = 12.454$ cal/deg-mole. At 300°K and 1 atm, $s = 22.985$ eu/mole.

25-19. Given the virial expansion $p = RT\rho + BRT\rho^2 + CRT\rho^3 + DRT\rho^4$, calculate the density expansion of μ in the form $\mu(\rho) = \mu(\rho_0) + RT \ln (\rho/\rho_0) +$ an expansion in powers of ρ, where ρ_0 is an arbitrary very small density for which the gas is ideal.

25-20. A thermodynamics text states that biphenyl melts at 343°K with Δh of 4400 cal/mole, that the solid has $c_p = 55.00$ cal/mole-deg and the liquid has $c_p = 63.80$ cal/mole-deg. It then derives the following expression for the heat of fusion as function of T: $\Delta h = 1381.6 + 8.80T$, in cal/mole. What should the answer be, and what did the text do wrong?

26 INTRODUCTION TO MULTICOMPONENT SYSTEMS

INTRODUCTION

So far the applications of thermodynamics treated in this book have been only to systems of fixed composition. The reason for this is simplicity; the properties of systems viewed as functions of composition are far more complicated than those of systems with fixed composition. Of course, the properties of multicomponent systems are correspondingly more interesting than those of single-component systems. The problems of chemistry, biology, geology, oceanography, engineering, pollution control, etc., always involve several components and the interactions between them. The best reference for the material of Secs. 26 through 29 of which the author is aware, and one to which he is indebted, is J. M. Prausnitz.*

A homogeneous substance containing two or more pure substances which are not chemically bound to each other is called a **solution**. Sometimes the word *mixture* is used to mean solution, but mixture has a broader application, since one talks about mixtures of materials that are immiscible and thus form heterogeneous systems. The definition of a solution is arbitrary in that the distinction between chemical bonds and the kinds of bonds holding liquids or solids together often fades away as one considers weaker and weaker

* J. M. Prausnitz, *Molecular Thermodynamics of Fluid-Phase Equilibria*, Prentice-Hall, Englewood Cliffs, N.J., 1969.

chemical bonds.   Also, there is arbitrariness in the definition of a pure substance; for example, do different isotopes of the same element or atoms with different nuclear spins give rise to the same or to different substances? We can answer by saying that if one is restricted to experiments in which such nuclear properties as mass and nuclear spin cannot be distinguished, then different isotopes or spins can be ignored consistently.   It is not always obvious whether these nuclear properties will affect macroscopic behavior, however.   For example, ortho- and parahydrogen have different heat capacities, and uranium hexafluoride diffuses at different rates through barriers depending on the mass of the uranium atom.   Finally, the term *solution* is generally reserved for the case in which the pure substances are thoroughly mixed at the molecular level.   Clumps of molecules, otherwise homogeneously dispersed throughout a system, are called a **colloidal suspension.** The word **solute** is usually taken to mean the pure substance(s) in lower concentration in the solution, the word **solvent** to mean the substance in higher concentration.   This usage is arbitrary, however, and the words may be used in whatever way is most convenient.

THE GIBBS PHASE RULE

For fixed composition the p-V-T diagram resembles Fig. 18-1 with projection onto the p-T plane (phase diagram) resembling Fig. 25-1.   When composition becomes a variable, things get complicated indeed.   For example, it is not even immediately clear how many independent intensive variables there are in a multicomponent, multiphase system.

The **number of degrees of freedom** f in a system is defined to be the number of intensive variables used to characterize the system minus the number of relations or restrictions connecting them.   For example, a single-component single-phase system may be described by the intensive quantities p, T, and μ. However, these variables are related by Eq. 13-31: $\Omega(T, p, \mu) = 0$, so such a system has $f = 2$.   A single-component system specified as having two coexisting phases may be described by the intensive variables $T^{(1)}$, $T^{(2)}$, $p^{(1)}$, $p^{(2)}$, $\mu^{(1)}$, and $\mu^{(2)}$.   However, we saw in Sec. 15 that $T^{(1)} = T^{(2)}$ and $p^{(1)} = p^{(2)}$ at equilibrium.   Furthermore, each phase has its own relationship of the form of Eq. 13-31: $\Omega^{(1)}(T^{(1)}, p^{(1)}, \mu^{(1)}) = 0$; $\Omega^{(2)}(T^{(2)}, p^{(2)}, \mu^{(2)}) = 0$.   Also, since the two phases must coexist, $\mu^{(1)} = \mu^{(2)}$.   Thus the six intensive quantities are coupled by five independent relations, and there is only one degree of freedom for a single-component two-phase system.   This is manifested by the fact that, if either T or p is known for a two-phase system, the other variable may simply be read off of Fig. 25-1.   A single-component system specified as having three coexisting phases has $f = 0$;

there is a unique set of intensive variables that characterizes the triple point for such a system. The general derivation of the value of f follows:

If c is the number of components and π is the number of phases, the system may be described by the temperature and pressure in each phase and by the chemical potential of each component in each phase. The number of such variables is found by enumerating them,

$$
\begin{array}{cccc}
T^{(1)} & T^{(2)} & \cdots & T^{(\pi)} \\
p^{(1)} & p^{(2)} & \cdots & p^{(\pi)} \\
\mu_1^{(1)} & \mu_1^{(2)} & \cdots & \mu_1^{(\pi)} \\
\mu_2^{(1)} & \mu_2^{(2)} & \cdots & \mu_2^{(\pi)} \\
\vdots & & & \\
\mu_c^{(1)} & \mu_c^{(2)} & \cdots & \mu_c^{(\pi)}
\end{array}
$$

to be $(c + 2)\pi$.

The relations coupling them are the following:

$$
\begin{aligned}
& T^{(1)} = T^{(2)} = \cdots = T^{(\pi)}, \\
& p^{(1)} = p^{(2)} = \cdots = p^{(\pi)}, \\
& \Omega^{(1)}(T^{(1)}, p^{(1)}, \mu_i^{(1)}) = 0, \\
& \vdots \\
& \Omega^{(\pi)}(T^{(\pi)}, p^{(\pi)}, \mu_i^{(\pi)}) = 0, \\
& \mu_1^{(1)} = \mu_1^{(2)} = \cdots = \mu_1^{(\pi)} \\
& \vdots \\
& \mu_c^{(1)} = \mu_c^{(2)} = \cdots = \mu_c^{(\pi)}.
\end{aligned}
$$

The number of equations coupling the variables is the number of equal signs in the foregoing equations: $2(\pi - 1)$ for the T's and p's, π for the Ω's, and $c(\pi - 1)$ for the μ's. Thus f is simply

$$
f = (c + 2)\pi - 2(\pi - 1) - \pi - c(\pi - 1),
$$

$$
\boxed{f = c - \pi + 2.} \tag{26-1}
$$

This is the so-called **Gibbs phase rule.** In our definition of component we have ruled out chemical reactions and chemical restrictions of various kinds which complicate the definition of c. These are discussed briefly in Sec. 30.

MULTICOMPONENT EQUATIONS OF STATE

The equation of state for a mixture,

$$
p = p(V, T, n_1, n_2, \ldots, n_c) \tag{26-2}
$$

is, in general, very complicated; for most systems only limited experimental data are available. Various graphical means of expressing part of the information contained in Eq. 26-2 (with different combinations of variables held constant) have been employed. They will not be treated here. Binary (that is, two-component) solutions are the easiest, since the intensive variables T, p, and a single mole fraction describe the system (since $x_1 + x_2 = 1$). For binary solutions, it is common to plot the compressibility factor $z = pV/(n_1 + n_2)RT$ against x_1 with lines of different constant pressure (isobars), as shown in Fig. 26-1.

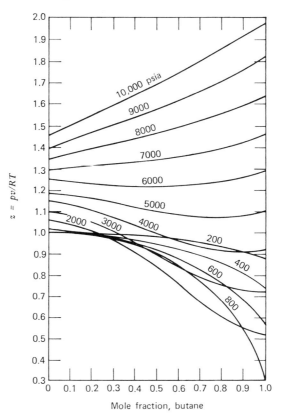

Fig. 26-1. Compressibility factor for mixtures of nitrogen and butane at 171°C. (*From* J. M. Prausnitz, *Molecular Thermodynamics of Fluid-Phase Equilibria*, Prentice-Hall, Englewood Cliffs, N.J., 1969, p. 33.)

The simplest analytic treatment of gas mixtures is the *ideal mixture of ideal gases*, for which each component contributes independently to the pressure:

$$p = \frac{(n_1 + n_2 + \cdots)RT}{V} \qquad \text{(ideal mixture).} \qquad (26\text{-}3)$$

The range of applicability of this approximation to noninteracting gases is similar to that for the single-component ideal gas law.

A slightly more complex treatment of fluid mixtures is given by **Amagat's law**, which says that at fixed temperature and total pressure the components mix with no change in total volume (that is, *isometrically*):

$$V(T, p, n_1, n_2, \ldots) = \sum_j n_j v'_j(T, p) \qquad \text{(isometric mixing),} \qquad (26\text{-}4)$$

where $v'_j(T, p)$ is the molar volume of pure j at the temperature and pressure of the solution. This implies that the partial molar volume of each component in the mixture is the same as its molar volume when pure:

$$v_j(T, p, x_1, x_2, \ldots) \equiv \left(\frac{\partial V}{\partial n_j}\right)_{T, p, n_k \neq i} = v'_j(T, p). \qquad (26\text{-}5)$$

In a sense, this means that the nonideality in the mixture is the same as the nonideality in the pure individual constituents at the same p and T. Insofar as the molecules in the mixture experience roughly the same intermolecular forces as they do in the pure state, Amagat's law is approximated. It thus is good at low pressures where the intermolecular forces can almost be neglected (Eq. 26-3 is a special case of Eq. 26-4). It is sometimes good at very high pressures where the hard-core volumes of the molecules predominate in establishing v_i, both for the pure and the mixed systems. It is also good when the molecules of the various components are similar, for example, with mixtures of Ar and O_2, of CO and N_2, or of compounds differing only isotopically. However, it is a poor approximation at moderate pressures when the molecular properties of the components are dissimilar.

It is often convenient to refer to the **partial pressure** p_j of a single component in a fluid mixture, defined by

$$\boxed{p_i \equiv y_j p,} \qquad (26\text{-}6)$$

where p is the total pressure in the fluid and y_j is the mole fraction of j. Clearly, the sum of the partial pressures of all the constituents yields p, and it is tempting to view the total pressure as made up of the sum of the separate effects of the molecules of each constituent. However, only for low-density gases, where the molecules do not interact appreciably with each other, is this interpretation accurate. For gas mixtures which depart significantly from ideality, the partial pressure must be viewed simply as a convenient mathematical construct.

A multicomponent fluid whose composition is fixed should obey the various analytic equations of state about as well as pure components do. Changing to a new composition is, then, equivalent to changing to a new fluid, and necessitates new values for the adjustable parameters in the equation. Therefore the main problem in using such equations is finding the concentration dependence of these adjustable parameters. There are no thermodynamic arguments to help, and one can resort only to experiment or to molecular theories. Often mixtures must be studied for which no experimental data are available; only the pure components have been studied. Then one must use molecular arguments and intuition to find how the values of the equation-of-state parameters change with concentration from those of the pure components. The rules governing this change are called *mixing rules*.

For example, with van der Waals' equation, molecular theories intimate a correspondence between the parameter a and the strength of the attractive forces between a pair of molecules, and between the parameter b and the size of the hard cores of the molecules. As a result, one might hope that b for a mixture would be some kind of average of the b's for the pure components. Several averages have been proposed, but the most commonly used mixing rule for van der Waals b, chosen for its simplicity, is

$$b(x_1, x_2, \ldots) = \sum_j x_j b_j. \tag{26-7}$$

For the parameter a, one might hope to average over all kinds of pairs of molecules:

$$a(x_1, x_2, \ldots) = \sum_{jk} x_j x_k a_{jk}, \tag{26-8}$$

where a_{jk} describes the interaction between one molecule of type j and one of type k. In this notation, a_{jj} is simply van der Waals' a for pure substance j. However, for $j \neq k$, in the absence of experimental data, a method is needed to relate the intermolecular potential of a j-k pair to the potentials of j-j and k-k pairs. Molecular theory can give no general answer to this problem. Common usage, backed up by some theoretical arguments, has settled on the *geometric mean* assumption:

$$a_{jk} = \sqrt{a_j a_k}, \tag{26-9}$$

Where one knows nothing else, the use of Eqs. 26-7, 26-8 and 26-9 with van der Waals' equation is at least far superior to the use of Amagat's law.

The same mixing rules have also been suggested for the a and b parameters of the Redlich-Kwong equation, based on similar molecular arguments. The parameters present in the modified Redlich-Kwong equation have also been assigned mixing rules, with a result giving fair to excellent accuracy.* The resulting equation is most reliable for mixtures of nonpolar components well away from the critical point. As illustration, Fig. 26-2 shows the calculated and experimental values of z for nitrogen-ethylene mixtures. However, for components whose molecules differ significantly in chemical nature or molecular size, there is no reason to trust any arbitrary mixing rules.

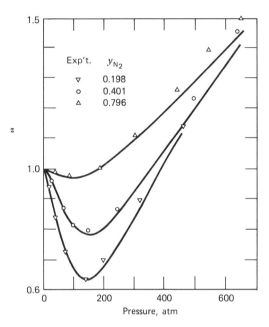

Fig. 26-2. Compressibility factor for mixtures of nitrogen and ethylene at 50°C, calculated from the modified Redlich-Kwong equation. (*From* O. Redlich, F. J. Ackerman, R. D. Gunn, M. Jacobson, and S. Lau, *I.&E.C. Fundamentals* **4**, 369 (1965).)

Use of the virial equation for gas mixtures has the advantage of a rigorous derivation from statistical mechanics. Thus one can write exact expressions for the concentration dependence of the coefficients,

$$B(x_1, x_2, \ldots) = \sum_{jk} x_j x_k B_{jk}, \tag{26-10}$$

$$C(x_1, x_2, \ldots) = \sum_{jkl} x_j x_k x_l C_{jkl}, \tag{26-11}$$

and all that is needed is to find the B_{jk} and C_{jkl} as functions of T from either

---

* O. Redlich, F. J. Ackerman, R. D. Gunn, M. Jacobson, and S. Lau, *I. & E. C. Fundamentals* **4**, 369 (1965). See also J. M. Prausnitz, *Molecular Thermodynamics of Fluid-Phase Equilibria*, Prentice-Hall, Englewood Cliffs, New Jersey, 1969, p. 153.

experiment or molecular theory. The B_{jj} and C_{jjj} are the second and third virial coefficients of pure j. Calculation of virial coefficients is limited in accuracy primarily because of insufficient understanding of intermolecular potential functions, especially among dissimilar molecules.

PROBLEMS

26-1. A mixture of 0.1 g H_2 and 0.2 g N_2 is to be stored at 760-mm pressure and 20°C. (*a*) What must the volume of the container be? (*b*) What is the mole fraction of H_2 in the mixture? (*c*) What is the partial pressure of H_2?

26-2. How many degrees of freedom has a mixture of sugar and water with one solid, one liquid, and one gaseous phase? What is meant by this result?

26-3. Prove that binary gas mixtures following Amagat's rule must have $B_{11} + B_{22}$ $= B_{12} + B_{21}$ (note: $B_{12} = B_{21}$ is always imposed with no loss in generality) $= 2B_{12}$.

26-4. The critical point of nitrogen is at $-147.1°C$, 33.5 atm, and 0.311 g/cm^3. That of ethylene (C_2H_4) is at 9.7°C, 50.9 atm, and 0.22 g/cm^3. If 0.4 mole of nitrogen is mixed with 0.6 mole of ethylene, and the mixture is contained in a 200-ml tank at 50°C, calculate p and z using van der Waals' equation. Compare your result with the experimental value of z for that pressure, shown in Fig. 26-2.

27   IDEAL SOLUTIONS

FUGACITY AND ACTIVITY

Each coexisting phase at equilibrium is a homogeneous solution. A goal of multicomponent thermodynamics is to find the relationships among the temperature, pressure, and equilibrium concentrations of each component in each phase. For example, in the distillation of a solution, what is the relation between the concentration of the vapor and that of the liquid? A variety of such problems in the absence of chemical reactions are studied in Secs. 27, 28, and 29; chemical reactions are treated in Sec. 30.

Since it is the value of the chemical potential μ_i of a substance which determines how that substance is dispersed among the phases at equilibrium, our task is to relate μ_i, an abstract mathematical quantity, to the temperature, pressure, and concentrations which are accessible to experiment. Of course, any such relationship must give μ_i *only with respect to its value in some arbi-*

trarily chosen reference state for substance i. This is because, on the one hand, in the equation $\mu_i = u_i - Ts_i + pv_i$ the zeros of u_i and s_i are arbitrary, or on the other hand, in the integration of the Gibbs-Duhem equation (for example, Eq. 25-5) to find μ_i, there appears an arbitrary constant of integration. The judicious choice of these so-called **standard states**, with respect to which μ_i is calculated, greatly facilitates the solution of particular problems in thermodynamics.

For a single-component system, changes in μ_i are given by Eq. 25-5:

$$d\mu_i = -s_i\, dT + v_i\, dp \qquad \text{(single component).} \qquad (27\text{-}1)$$

If substance i is an ideal gas at fixed temperature T, this may be integrated from a reference pressure p^0 to yield

$$\int d\mu_i = \int \frac{RT\, dp}{p} \qquad \text{(pure ideal gas, } T \text{ constant),} \qquad (27\text{-}2)$$

$$\boxed{\mu_i - \mu_i{}^0 = RT \ln \frac{p}{p^0}} \qquad \text{(pure ideal gas, } T \text{ constant).} \qquad (27\text{-}3)$$

This represents for one simple substance a relationship between μ_i and the pressure, which is easily accessible to experiment. The quantity $\mu_i{}^0$ is simply the chemical potential of the pure ideal gas i at the pressure p^0, since $\ln (p^0/p^0) = 0$.

With Eq. 27-3 as motivation, G. N. Lewis defined a function called the **fugacity** of sustance i, f_i, by the equation

$$\boxed{\mu_i(T, p, x_1, x_2, \ldots) - \mu_i{}^0 \equiv R T \ln \frac{f_i}{f_i{}^0}} \qquad \text{(any system, } T \text{ constant),}$$
$$(27\text{-}4)$$

which holds for any component of any system. Here, $\mu_i{}^0$ is the chemical potential of substance i in the state with fugacity $f_i{}^0$. It is often convenient, especially for gases, to choose as the standard state the gas in question at temperature T *in the hypothetical ideal-gas condition* at 1 atm of pressure. In this case, $f_i{}^0$ is 1 atm, and the causes of departures of f_i from the value 1 atm are due to changes in the partial pressure of i from 1 atm and to non-ideality of the gas mixture. For a pure ideal gas, $f_i = p_i$; similarly, in a mixture of ideal gases it equals the partial pressure, $f_i = p_i = y_i p$. (For convenience, we will use y_i for mole fraction of i in a gaseous phase and x_i for mole fraction of i in a condensed phase.) Thus in the limit of vanishing density, f_i defined by Eq. 27-4 must become

$$\lim_{\rho \to 0} f_i = y_i p \qquad \text{(any system).} \qquad (27\text{-}5)$$

Therefore, fugacities have the units of pressure (usually atmospheres). They may be thought of as "corrected pressures" or "active pressures," a concept which aids one's physical intuition.

Lewis called the quantity (f_i/f_i^0) the **activity** a_i of substance i, that is,

$$\mu_i(T, p, x_1, x_2, \ldots) \equiv \mu_i^0 + RT \ln a_i \qquad \text{(any system, } T \text{ constant)}. \tag{27-6}$$

Thus activities are dimensionless. They are measures of how "active" the substance is compared to when it is in its standard state at the same temperature; thus they may be viewed as "modified or active concentrations." *An activity or fugacity by itself is meaningless unless one knows what standard state was used in its determination.*

If the two phases α and β are in equilibrium, then

$$\mu_i^\alpha = \mu_i^\beta, \tag{27-7}$$

$$\mu_i^{0\alpha} + RT \ln \frac{f_i^\alpha}{f_i^{0\alpha}} = \mu_i^{0\beta} + RT \ln \frac{f_i^\beta}{f_i^{0\beta}}, \tag{27-8}$$

where we have acknowledged the possibility of using a different standard state for sustance i in the two states. The relationship between $\mu_i^{0\alpha}$ and $\mu_i^{0\beta}$ may be found from Eq. 27-4 by letting μ_i^0 be $\mu_i^{0\beta}$ and μ_i be $\mu_i^{0\alpha}$:

$$\mu_i^{0\alpha} - \mu_i^{0\beta} = RT \ln \frac{f_i^{0\alpha}}{f_i^{0\beta}}. \tag{27-9}$$

Substitution of this into Eq. 27-8 yields

$$RT \ln \frac{f_i^\alpha}{f_i^\beta} = 0,$$

or

$$f_i^\alpha = f_i^\beta \qquad \text{(phase equilibrium)}. \tag{27-10}$$

Thus regardless of whether the same standard state is chosen for substance i in the phases, the fugacity of i will be equal in all phases at equilibrium, that is, Eq. 27-10 is completely equivalent to Eq. 27-7. It is clear from the definition, Eq. 27-6, of activities that $a_i^\alpha = a_i^\beta$ only when the same standard state is used for i in the two phases. It now becomes our task to relate fugacities or activities of constituents of various solutions to accessible quantities like T, p, and concentrations.

IDEAL SOLUTIONS AND RAOULT'S LAW

An *ideal solution* is one for which the fugacity of each component is simply proportional to the fraction of the molecules in the solution which are of that component, that is, to the mole fraction of the component:

$$f_i(T, p, x_i) = x_i f_{\text{pure } i}(T, p) \qquad \text{(ideal solution, for all } i\text{).} \qquad (27\text{-}11)$$

The relationship defines ideal solid solutions, liquid solutions, or gaseous solutions (where y_i would be used in place of x_i). Thus if pure i at T and p is used as the standard state for substance i, the chemical potential of i in an ideal solution is

$$\mu_i = \mu_i{}^0 + RT \ln x_i \qquad \text{(ideal solution, for all } i\text{),} \qquad (27\text{-}12)$$

that is, the activity of i is simply its mole fraction.

Now suppose that a condensed phase which is an ideal solution is in equilibrium with a vapor phase, which is also an ideal solution, at T and p. For each constituent,

$$f_i^c(T, p, x_i) = f_i^v(T, p, y_i), \qquad (27\text{-}13)$$

$$x_i f^c_{\text{pure } i}(T, p) = y_i f^v_{\text{pure } i}(T, p), \qquad (27\text{-}14)$$

where Eq. 27-13 is simply Eq. 27-10, and Eq. 27-14 is simply recognition of the ideality of the two solutions. We saw in Sec. 25 that the chemical potential of a condensed phase changes very slowly with pressure, so to a reasonable approximation we can replace

$$f^c_{\text{pure } i}(T, p) \approx f^c_{\text{pure } i}(T, p_i{}^0), \qquad (27\text{-}15)$$

where $p_i{}^0$ is the vapor pressure of pure condensed i at the temperature T. Now the equilibrium at T between pure condensed i and its vapor yields

$$f^c_{\text{pure } i}(T, p_i{}^0) = f^v_{\text{pure } i}(T, p_i{}^0), \qquad (27\text{-}16)$$

so, finally, Eq. 27-14 becomes

$$x_i f^v_{\text{pure } i}(T, p_i{}^0) = y_i f^v_{\text{pure } i}(T, p), \qquad (27\text{-}17)$$

where Eqs. 27-15 and 27-16 were used. Now if pure i vapor at T and at both $p_i{}^0$ and p is an ideal gas, each fugacity in Eq. 27-17 may be replaced by the pressure:

$$\boxed{x_i p_i{}^0(T) = y_i p \equiv p_i} \qquad \text{(2 ideal solutions, Eq. 27-15, pure } i \text{ gas is ideal)}$$

$$(27\text{-}18)$$

which is called **Raoult's law**. It says that the partial vapor pressure of

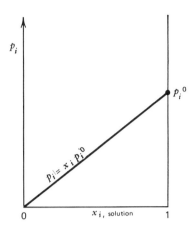

Fig. 27-1. Vapor pressure curve for one component obeying Raoult's law.

component i is simply proportional to the fraction of the molecules in the condensed phase that are i. This is shown graphically in Fig. 27-1, a plot of the equilibrium partial pressure of i in the vapor above a condensed solution at some fixed temperature. At $x_i = 0$, there is, of course, no pressure of i in the vapor because there is no i in the system.   As x_i increases and the probability of i molecules lying near the surface increases, p_i increases linearly.   At $x_i = 1$, the system is pure i, and the p_i becomes p_i^0, which is the vapor pressure of pure substance i at the temperature concerned.   As the temperature is increased, p_i^0 is increased, so Fig. 27-1 would assume a steeper and steeper slope for increasing T.

Note: The expression "ideal solution" is used in a variety of ways.   Often it is used for a solution each component of which obeys Raoult's law.   Some authors of standardized examinations assume that everyone defines it this way.   Indeed, if the total vapor pressure is not too high, the only severe assumption in the derivation of Raoult's law is that the condensed solution be ideal, so the statements are not too different.   The phrase also is applied to solutions in the dilute concentration range in which Henry's law (see Sec. 28) is obeyed.

A number of solutions, especially those in which the molecules of the constituents are very much alike, do approximate ideality.   Benzene and toluene are examples, and vapor pressure curves for them are shown in Fig. 27-2 for the constant temperature 60°C.   The partial pressures of benzene and toluene, each of which obeys Eq. 27-18, are shown (where, of course, $x_{ben} + x_{tol} = 1$), plus the total vapor pressure curve, which is just the sum of the partial pressures.

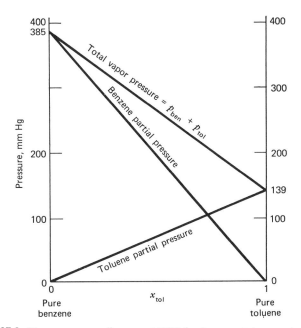

Fig. 27-2. Vapor pressure diagram at 60°C for benzene-toluene solutions.

As Eq. 27-18 indicates, the mole fraction y_i of i in the vapor is, in general, different from its value x_i in the condensed phase, a fact employed in the use of distillation to change the composition of materials. Since y_i is just the pressure fraction in the vapor, its value may be found immediately from Eq. 27-18 for solutions obeying Raoult's law:

$$y_i \equiv \frac{p_i}{p} = \frac{x_i \, p_i{}^0}{\sum_k x_k \, p_k{}^0} \qquad \text{(Raoult's law)}. \qquad (27\text{-}19)$$

Example 27-1. Derive the numerical formula relating y_{tol} to x_{tol} for mixtures of toluene and benzene at 60°C. If x_{tol} is 0.60, what is the partial vapor pressure of each component, the total vapor pressure, and y_{tol}?

Solution: The general formula may be derived from Eq. 27-19 by substituting the values of $p_k{}^0$ from Fig. 27-2:

$$y_{tol} = \frac{x_{tol}(139)}{x_{tol}(139) + x_{ben}(385)} = \frac{x_{tol}(139)}{x_{tol}(139) + (1 - x_{tol})(385)},$$

$$y_{tol} = \frac{139 x_{tol}}{385 - 246 x_{tol}}. \qquad (27\text{-}20)$$

The partial vapor pressures may be found directly from Raoult's law:

$$p_{tol} = x_{tol} \, p_{tol}^0 = (0.6)(139 \text{ mm}) = 83.4 \text{ mm}, \qquad (27\text{-}21)$$

$$p_{ben} = x_{ben} \, p_{ben}^0 = (0.4)(385 \text{ mm}) = 154 \text{ mm}, \qquad (27\text{-}22)$$

$$p = p_{tol} + p_{ben} = 83.4 \text{ mm} + 154 \text{ mm} = 237.4 \text{ mm}, \qquad (27\text{-}23)$$

$$y_{tol} = \frac{p_{tol}}{p} = \frac{83.4 \text{ mm}}{237.4 \text{ mm}} = 0.351. \qquad (27\text{-}24)$$

Hopefully, the formulas used seem more intuitively sensible than simply memorized.

Figure 27-3 is the distillation isotherm for benzene-toluene at 60°C. The straight line in Fig. 27-3 is simply the top line of Fig. 27-2, that is, the total vapor pressure of the solution as a function of x_{tol}.   Also plotted is

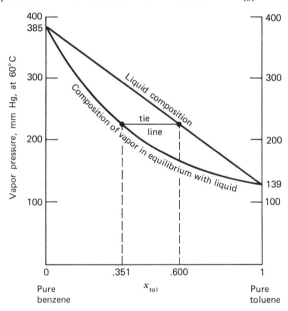

Fig. 27-3. Total vapor pressure curve at 60°C for benzene-toluene solutions with appropriate vapor composition curve.

a line giving the composition of the vapor which exerts that pressure in equilibrium with that liquid.   The space between the curves is filled with horizontal tie lines the ends of which give the compositions of the liquid and vapor which are in equilibrium with each other at that total pressure.   The particular tie line indicated in Fig. 27-3 is that which was computed in Example 27-1.

Of the two components in a binary solution, the **more volatile** is the one with the higher vapor pressure when pure. With minor exceptions in the cases of solutions where azeotropes form (see Sec. 28), *the vapor is always richer than the liquid in the more volatile constituent.* This statement can be proved for ideal binary solutions by deriving, as is done in Prob. 27-13, the result

$$\frac{y_1}{x_i} = \frac{1}{1 + (D/p_1{}^0)x_2},$$
(27-25)

where

$$D \equiv p_2{}^0 - p_1{}^0.$$
(27-26)

If D is positive, this means from Eq. 27-26 that substance 2 is more volatile than substance 1. It also means from Eq. 27-25 that the vapor is poorer in 1 and richer in 2 than the liquid. Similarly, if D is negative, then 1 is more volatile than 2, and the vapor is richer in 1 and poorer in 2 than the liquid. It is well to think of the vapor line in Fig. 27-3 as lying *to the left* of the liquid line rather than underneath it, since the tie lines are horizontal. Because benzene is more volatile than toluene, the vapor is richer in benzene than the liquid, so the vapor line must lie to the left of the liquid line.

If a material is stable in the liquid (or solid) state, it is the pressure exerted on it which maintains that stability. No condensed phase is stable at low enough pressure, because at any finite temperature molecules can occasionally get enough energy to break free of the bulk and leave. If the pressure is low enough they will not return. Of course, the times required for many solids to sublime may be very long, especially at low temperatures. Gravitational attraction is also important for very large masses in keeping particles from escaping. This is lucky for the inhabitants of planets, since interstellar space is a *very*-low-pressure gas!

When the vapor pressure of a liquid reaches the pressure applied to that liquid, the liquid structure is no longer stable and the liquid boils. Thus the **boiling point** is the temperature at which the vapor pressure of a liquid equals the applied pressure on the liquid. The **standard boiling point**, so-called, is the temperature at which the vapor pressure of a liquid equals 760 mm.

We can think of the boiling of benzene-toluene solutions as follows: At 60°C, Fig. 27-3 shows the vapor pressures of various solutions. Neither pure substance boils at so low a temperature, thus no ideal solution of them does. If the temperature is raised, the vapor pressures of both pure substances increase in accordance with the Clapeyron equation, Eq. 25-10 (or Clausius-Clapeyron equation). Since benzene is more volatile than toluene, the vapor pressure of benzene reaches 760 mm at a lower temperature than does that of toluene.   Benzene, it happens, boils at 80°C and toluene at 111°C.   Ideal solutions of the two boil at intermediate temperatures. Figure 27-4 shows a plot of boiling temperatures of mixtures of various compositions.   To the left of the liquid line, connected by horizontal tie lines, is plotted the composition of the vapor which is in equilibrium with the liquid that is boiling at the particular temperature.   The pressure is fixed in Fig. 27-4 at 760 mm.

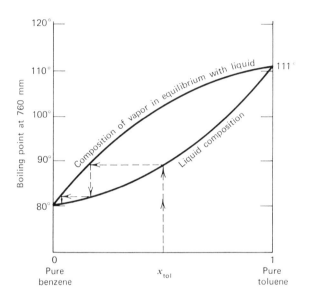

Fig. 27-4. Boiling point diagram for benzene-toluene solutions.

Suppose a benzene-toluene solution of $x_{tol} = \frac{1}{2}$ was heated in a pot at 1 atm of pressure.   The temperature moves up the dashed line in Fig. 27-4, and at approximately 89°C, as shown in the figure, the solution will boil. The composition of the vapor in equilibrium with that boiling liquid is found from the other end of the horizontal tie line to be about $x_{tol} = 0.2$.   If that vapor were cooled until it condensed to a liquid, following the dashed line downward, *its* boiling point would be about 82°C, and the vapor in equilibrium with it as it boiled would have x_{tol} approximately equal to 0.05.   The procedure of **fractional distillation** is one of simply repeating this boiling and condensing a number of times.   As one goes up the fractional distillation column to cooler temperatures where the material has been vaporized and condensed several times, the concentration is progressively enriched in the more volatile component, in this case in benzene.   If the fractionating column is long and efficient, the substance coming off at the top is pure benzene. As benzene is removed, the stuff in the pot must get richer in toluene, and so as the distillation proceeds, the point representing pot material moves to the right on the diagram. One can continuously collect benzene from the top, thus enriching the pot material in toluene, until finally an almost complete separation of benzene and toluene is accomplished.

To understand the lowering of the vapor pressure of solutions as non-volatile solutes (sugars or salts, for example) are dissolved in volatile solvents involves direct application of Eq. 27-18 and Fig. 27-1. Pure solvent has the vapor pressure $p_1{}^0$. A **nonvolatile** substance has zero vapor pressure, for all practical purposes. If their solutions obey Raoult's law,

$$p = p_1 = x_1 p_1{}^0 = (1 - x_2) p_1{}^0$$

(Raoult's law, component 2 nonvolatile). (27-27)

This is also the total vapor pressure of the solution. The specific nature of the nonvolatile component 2 is unimportant, only its mole fraction matters. If substance 2 dissociates in solution to form two or more particles (such as ions), each particle contributes to the mole fraction x_2 of nonvolatile solute.

FORMATION OF IDEAL SOLUTIONS

By focusing briefly on the process of forming ideal solutions, we gain insight into the nature of the ideal-solution approximation. Each component in an ideal solution has its chemical potential given by

$$\mu_i(T, p, x_1, x_2, \ldots) = \mu_i{}^0(T, p) + RT \ln x_i \quad \text{(ideal solution),} \quad (27\text{-}28)$$

where pure i at the temperature T and pressure p of the system is the standard state. We presume the solution is ideal over a range of values of T and p, and differentiate Eq. 27-28 with respect to p:

$$\left(\frac{\partial \mu_i}{\partial p}\right)_{T, \text{conc}} = \left(\frac{\partial \mu_i{}^0}{\partial p}\right)_T \quad \text{(ideal solution).} \quad (27\text{-}29)$$

The values of the derivatives may be found by considering the equation of the form of Eq. 27-1 for multicomponent systems:

$$d\mu_i = -s_i \, dT + v_i \, dp + \sum_j \left(\frac{\partial \mu_i}{\partial x_j}\right)_{T, p, x_{k \neq j}} dx_j. \quad (27\text{-}30)$$

This is a general equation relating the change in μ_i to explicit changes in T, p, and the mole fractions. Of course, there are only $c - 1$ independent mole fractions (where c is the number of components), because the sum of all the

x_i's must be unity. From Eq. 27-30 the pressure derivatives in Eq. 27-29 are seen to be the partial molar volumes:

$$v_i(T, p, x_i, x_2, \ldots,) = v_i^0(T, p) \qquad \text{(ideal solution).} \qquad (27\text{-}31)$$

This says that the partial molar volume of i in the ideal solution equals the molar volume of pure i at the same temperature and at a pressure equal to the total pressure on the solution. This is simply Amagat's law, Eq. 26-4. Since large systems have (see Eq. 12-13) a total volume $V = \sum_j n_j v_j$, Eq. 27-31 means that the mixing process which forms ideal solutions, if done at constant temperature and total pressure, occurs at constant total volume:

$$\Delta V_{\text{mix}} = V_{\text{final}} - V_{\text{init}} = \sum_j n_j(v_j - v_j^0) = 0 \quad \text{(ideal solution).} \quad (27\text{-}32)$$

This is shown schematically in Fig. 27-5. It is noteworthy that the partial pressures of the components change from p down to p_i during this isometric mixing.

| n_1 of 1
 T
 p
 V_1^0 | $+$ | n_2 of 2
 T
 p
 V_2^0 | $\rightarrow$ | n_1 of 1 and n_2 of 2
 T
 $p = p_1 + p_2$
 $V = V_1^0 + V_2^0$ |

Fig. 27-5. Isometric mixing process at constant T and total pressure.

Next let us divide Eq. 27-28 through by T and differentiate the result with respect to T:

$$\left[\frac{\partial(\mu_i/T)}{\partial T} \right]_{p, \text{conc}} = \left[\frac{\partial(\mu_i^0/T)}{\partial T} \right]_p \qquad \text{(ideal solution).} \qquad (27\text{-}33)$$

The values of the derivatives may be found most conveniently by writing μ_i as the partial molar Gibbs free energy:

$$\mu_i = h_i - T s_i, \qquad (27\text{-}34)$$

$$\frac{\mu_i}{T} = \frac{h_i}{T} - s_i, \qquad (27\text{-}35)$$

$$\left[\frac{\partial(\mu_i/T)}{\partial T} \right]_{p, \text{conc}} = -\frac{h_i}{T^2} + \frac{1}{T}\left(\frac{\partial h_i}{\partial T} \right)_{p, \text{conc}} - \left(\frac{\partial s_i}{\partial T} \right)_{p, \text{conc}}. \qquad (27\text{-}36)$$

The last two terms in Eq. 27-36 are seen from the Gibbs equation for enthalpy.

$$dh_i = T\, ds_i + v_i\, dp \qquad \text{(concentration constant)} \qquad (27\text{-}37)$$

to be the same, but of opposite sign (this is analogous to Eq. 19-6). Thus

$$\left[\frac{\partial(\mu_i/T)}{\partial T}\right]_{p,\text{conc}} = -\frac{h_i}{T^2}, \qquad (27\text{-}38)$$

a general result, often called the **Gibbs-Helmholtz equation**, which has a number of applications. When Eq. 27-38 is used in Eq. 27-33 for ideal solutions, the result is

$$h_i(T, p, x_1, x_2, \ldots) = h_i^0(T, p) \qquad \text{(ideal solution)}. \qquad (27\text{-}39)$$

The partial molar enthalpy of i in the ideal solution equals the molar enthalpy of pure i at the same temperature and at a pressure equal to the total pressure on the solution. This means that the isometric mixing process of Fig. 27-5 occurs at constant enthalpy for ideal solutions,

$$\Delta H_{\text{mix}} = H_{\text{final}} - H_{\text{init}} = \sum_j n_j(h_j - h_j^0) = 0 \qquad \text{(ideal solution)}. \quad (27\text{-}40)$$

Since the heat of a constant pressure process is ΔH for that process, the formation of an ideal solution from pure constituents at constant temperature and total pressure occurs with no heat effect. The so-called *heat of mixing*, ΔH_{mix}, *to form an ideal solution is zero.*

Next, let us differentiate Eq. 27-28 with respect to T:

$$\left(\frac{\partial \mu_i}{\partial T}\right)_{p,\text{conc}} = \left(\frac{\partial \mu_i^0}{\partial T}\right)_p + R \ln x_i \qquad \text{(ideal solution)}. \qquad (27\text{-}41)$$

The temperature derivatives are identified immediately from Eq. 27-30:

$$-s_i(T, p, x_1, x_2, \ldots) = -s_i^0(T, p) + R \ln x_i \qquad \text{(ideal solution)}. \quad (27\text{-}42)$$

The partial molar entropy of i in the ideal solution equals the molar entropy of pure i at the same temperature and at a pressure equal to the total pressure on the solution, less the term $R \ln x_i$. The entropy change for the mixing process of Fig. 27-5 for ideal solutions is

$$\Delta S_{\text{mix}} = S_{\text{final}} - S_{\text{init}} = \sum_j n_j(s_j - s_j^0) = -R \sum_j n_j \ln x_j \qquad \text{(ideal solution)}.$$
$$(27\text{-}43)$$

Since mole fractions are always less than unity, the logarithm terms are all negative. Thus the "entropy of mixing" at constant T and total pressure to form an ideal solution is positive. This makes sense, because such mixing is an irreversible adiabatic process. The spontaneous unmixing in which the right-hand side of Fig. 27-5 would suddenly find all the molecules of type 1 on the left and all of type 2 on the right is inconceivable. Such spontaneous ordering of an isolated portion of the universe has never been observed, as noted in our statement of the second law.

One final comment on the formation of ideal solutions is to note that μ_i in Eq. 27-28 shows no dependence on the other constituents of the solution. The sole concentration on which μ_i depends is x_i; $\partial\mu_i/\partial x_j = 0$ as long as x_i is kept constant (which it cannot be in binary solutions, since $x_1 + x_2 = 1$ or $dx_1 = -dx_2$ for them).

MIXING OF IDEAL GASES

There is a sense in which the expression "entropy of mixing," at least for ideal mixtures of ideal gases, is a misnomer. If ideal gases were mixed *reversibly*, as discussed conceptually in Sec. 12 (see Fig. 12-1), the entropy change would be $\int dQ/T$, and since no heat is required to restore the temperature to its original value, this is zero. The *reversible* mixing, however, is a process at constant chemical potential; thus for ideal gases (as shown in Eqs. 27-3 to 27-5) it is a process at constant *partial* pressure (or, equivalently, constant volume available to each gas) and not total pressure. The entropy increase of Eq. 27-43 for ideal-gas mixtures comes from the fact that every constituent gas finds its pressure decreased, its volume increased, in what is otherwise simply a constant-temperature expansion. The molecules of one component do not even know the molecules of the others are present in the final mixture.

Since one has common occasion to calculate ΔS for the mixing or unmixing of ideal gases, it is well to develop a standard approach for such problems. Suppose one is given the problem of Fig. 27-6, with the data on initial and final states given as volumes rather than pressures. One may view the

$$
\boxed{\begin{array}{c} n_1 \text{ of } 1 \\ V_1 \\ T \end{array}} + \boxed{\begin{array}{c} n_2 \text{ of } 2 \\ V_2 \\ T \end{array}} \rightarrow \boxed{\begin{array}{c} n_1 \text{ of } 1 \text{ and } n_2 \text{ of } 2 \\ V_f \\ T \end{array}}
$$

Fig. 27-6. Initial and final states of a gas-mixing process.

changes which actually occur in this process as (1) the isothermal change of volume for n_1 moles of 1 from V_1 to V_f, and (2) the isothermal change of volume for n_2 moles of 2 from V_2 to V_f. The fact that we choose to store the two gases in *the same* container of volume V_f is immaterial, since the gas molecules do not interact. Calculation of ΔS_1 and ΔS_2 for these isothermal volume changes is immediate (see Sec. 21):

$$
dS = \left(\frac{\partial S}{\partial V}\right)_T dV = \left(\frac{\partial p}{\partial T}\right)_V dV \qquad (T \text{ constant}), \qquad (27\text{-}44)
$$

where the derivative was Maxwellized. On using the ideal-gas law this becomes

$$dS = \frac{nR\,dV}{V} \quad (T \text{ constant, ideal gas}). \quad (27\text{-}45)$$

Integration of Eq. 27-45 for the volume change $V_1 \to V_f$ for gas 1 and $V_2 \to V_f$ for gas 2 yields an entropy change for the mixing process of

$$\boxed{\Delta S = n_1\,R\ln\frac{V_f}{V_1} + n_2\,R\ln\frac{V_f}{V_2}} \quad \text{(mixing ideal gases, } T \text{ constant).}$$
$$(27\text{-}46)$$

If partial pressures are more convenient independent variables than volumes, the ideal-gas law ($p_i V_i =$ constant at constant T) may be used with Eq. 27-46 to yield

$$\Delta S = n_1 R\ln\frac{p_{1,\,\text{init}}}{p_{1,\,\text{final}}} + n_2\,R\ln\frac{p_{2,\,\text{init}}}{p_{2,\,\text{final}}}. \quad (27\text{-}47)$$

It must be emphasized in Eq. 27-47 that the final pressures of 1 and 2 are the final *partial* pressures, not the total final pressure.

PROBLEMS

27-1. Calculate the composition of a benzene-toluene solution which boils at 1 atm and 90°C, assuming an ideal solution. Also calculate the mole fraction of benzene in the vapor. At 90°C, benzene has a vapor pressure of 1022 mm when pure, and pure toluene has a vapor pressure of 406 mm. Use a sketch to aid visualization.

27-2. A company takes 1,000,000 (10^6) l of air (1 atm, 25°C), separates the oxygen and nitrogen, and bottles each separately at 50 atm. Calculate ΔS for the gas (treated as ideal; air 20% oxygen, 80% nitrogen by volume).

27-3. Ethyl and methyl alcohol form nearly ideal solutions. At 20°C, the vapor pressures of these compounds are 44.5 and 88.7 torr, respectively (1 torr = 1 mm Hg). If 100 g of each are mixed, find the partial vapor pressure of each gas and the total vapor pressure of the solution. What is the mole fraction of methanol in the vapor? Use a sketch to aid visualization.

27-4. A and B form ideal solutions. The vapor pressure of pure A is 950 mm, of pure B is 210 mm. What is the mole fraction of A in the solution which just boils at 1 atm of pressure? What is the mole fraction of A in the vapor that boils off? Use a sketch to aid visualization.

27-5. Gas for use in rescue work could be made as follows: Ten moles of N_2 originally at 50 atm of pressure and twenty moles of O_2 originally at 60 atm of pressure are put into a cylinder in which the final total pressure is 20 atm. Temperature is constant at 25°C. Treat as ideal gases, which originally were pure and, finally, were mixed in the cylinder. Calculate ΔU, ΔH, ΔS, ΔA, and ΔG.

27-6. A 1-l bulb containing nitrogen at 1.0 atm of pressure and 25°C is connected by a tube with a stopcock to a 3.0-l bulb containing carbon dioxide at 2.0 atm of pressure and 25°C. The stopcock is opened, and the gases are allowed to mix until equilibrium is reached. Assume that the gases are both ideal: (a) What is ΔS for this spontaneous change? (b) How would that ΔS be changed if the gas were heated to 150°C? *Note*: CO_2 is a linear molecule.

27-7. What is the entropy change in taking a quantity of air containing 1 mole of gas and separating it into pure nitrogen and pure oxygen? Total pressure of air is 1 atm, and final pressures of pure gases are 1 atm each. Temperature is constant at 30°C.

27-8. What is ΔG for the process of Prob. 21-21? *Hint*: A simple approach is to prove

$$\left[\frac{\partial(\Delta\mu_i/T)}{\partial(1/T)}\right]_p = \Delta h_i \qquad (27\text{-}48)$$

from Eq. 27-38 (see footnotes to Eq. 30-34) and integrate.

27-9. A and B form ideal solutions. Their vapor pressures when pure are 300 and 500 mm respectively at 27°C. Their heats of vaporization are 10 and 15 kcal/mole respectively. Prove that the boiling point of a 50 mole percent solution of A and B at 750 mm pressure is given by T_2 in the following equation: 3 exp $[16.67(1 - 300/T_2)] + 5$ exp $[25(1 - 300/T_2)] = 15$.

27-10. The molecular weight of mannitol is 182.2. If 54.1 g of mannitol is dissolved in 1000 g of water at 20°C, by how much would the vapor pressure be lowered from its original value of 17.54 mm if the solution was ideal? If the same amount of a substance with the same molecular weight as mannitol were dissolved in 1000 g of water, and the substance was completely ionized into two ions per molecule, how would the above result be changed?

27-11. The vapor pressure of a solution of g_2 g of nonvolatile compound 2 in g_1 g of compound A is p. The vapor pressure of pure A at that temperature is p^0. A has molecular weight M_A and the solution is ideal. What is M_B?

27-12. What is the ratio between the fugacity of pure benzene and the fugacity of benzene in a solution containing 20 g of benzene, 10 g of diethyl ether, and 15 g of carbon bisulfide. Assume an ideal solution.

27-13. Prove Eq. 27-25.

27-14. One often reads in thermodynamics texts that if a solution is formed isometrically and with no heat of mixing (that is, Eq. 27-31 and Eq. 27-39 are valid), then the solution must be ideal (that is, Eq. 27-28 must be valid). From this would follow the validity of Eq. 27-42. Prove that Eq. 27-28 follows from the validity of Eqs. 27-31 and 27-39 only if the isometric and isenthalpic mixing occurs for the solution of interest at *all* pressures from the very lowest (where all constituents are essentially ideal gases) up to the pressure of interest.

28 REAL SOLUTIONS

FUGACITY AND ACTIVITY COEFFICIENTS

The ideal solution is a good approximation for mixtures of gases at low density, for dilute solutions, and sometimes for mixtures of liquids whose molecules are very similar. For other kinds of solutions, however, some consideration of the effects of nonideality must be made. It is convenient to retain the ideal solution as a basis of comparison from which to measure deviations. Thus the *fugacity coefficient* φ_i is defined (most commonly for gases) by

$$f_i \equiv \varphi_i y_i p \equiv \varphi_i p_i, \tag{28-1}$$

where y_i is the mole fraction of i, p is the total pressure on the solution, and f_i is the fugacity of i relative to a particular choice of standard state as defined in Eq. 27-4. With the usual choice of standard state, mixtures of ideal gases have $f_i = y_i p$, thus they have $\varphi_i = 1$. In this case, deviations from unity in the fugacity coefficient measure departures from ideality in the gas mixture.

Similarly, the *activity coefficient* γ_i is defined (most commonly for liquids or perhaps solids) by

$$f_i \equiv \gamma_i x_i f_i^0 \qquad \text{or} \qquad a_i \equiv \gamma_i x_i, \tag{28-2}$$

or, equivalently, by

$$\mu_i(T, p, x_1, x_2, \ldots) \equiv \mu_i^0 + RT \ln \gamma_i x_i. \tag{28-3}$$

The activity coefficient (or fugacity coefficient) for a real solution is a function of T, p, concentration of all constituents (not just of substance i), and of choice of standard state. Standard states are usually chosen so that over the most convenient range of concentrations the values of γ_i are nearly unity. The temperature in the standard state is always taken to be that of the solution.

At fixed temperature, volumetric measurements alone permit complete specification of all the state functions. Let us find the chemical potential relative to the standard state of an ideal gas at 1 atm of pressure by starting with $\mu_i{}^0$ for the substance i in the standard state and lowering the pressure on this ideal gas to nearly zero. We then mix up a solution to the desired composition, all gases being ideal. Finally, we increase the pressure on the solution to the desired value p. The whole procedure is at constant T, and $\Delta\mu_i$ may be calculated for each step.

In the standard state,

$$\mu_i{}^0 = h_i{}^0 - Ts_i{}^0 \qquad \text{(ideal gas, standard state).} \qquad (28\text{-}4)$$

Reducing the pressure (by increasing the volume) at constant temperature to a value ε which is almost zero changes μ_i by

$$\Delta\mu_i = \int_{1\,\text{atm}}^{\varepsilon} \left(\frac{\partial\mu_i{}^0}{\partial p}\right)_T dp = \int_{1\,\text{atm}}^{\varepsilon} v_i{}^0 \, dp = \int_{1\,\text{atm}}^{\varepsilon} \frac{RT}{p} \, dp. \qquad (28\text{-}5)$$

At this low pressure, all constituents of the real solution are ideal gases, and we mix them at constant total pressure ε. As was seen in Sec. 27, this leaves the enthalpy unchanged (Eq. 27-39) and changes the entropy (Eq. 27-42) by $-R \ln y_i$, where y_i is the mole fraction of i in the solution. Thus the chemical potential change for this zero-pressure mixing is

$$\Delta\mu_i = -T \, \Delta s_i = RT \ln y_i. \qquad (28\text{-}6)$$

Now the pressure can be increased on the real fluid mixture at constant temperature and composition to its final value p, which changes μ_i by

$$\Delta\mu_i = \int_{\varepsilon}^{p} \left(\frac{\partial\mu_i}{\partial p}\right)_T dp = \int_{\varepsilon}^{p} v_i \, dp. \qquad (28\text{-}7)$$

We now have found the total change, Eqs. 28-5 through 28-7, in μ_i on taking substance i from its standard state to its final state in the solution:

$$\mu_i = \mu_i{}^0 + \int_{1\,\text{atm}}^{\varepsilon} \frac{RT}{p} \, dp + RT \ln y_i + \int_{\varepsilon}^{p} v_i \, dp. \qquad (28\text{-}8)$$

It is convenient to simplify this result by rewriting the first integral:

$$\int_{1\,\text{atm}}^{\varepsilon} \frac{RT}{p} \, dp = -\int_{\varepsilon}^{1\,\text{atm}} \frac{RT}{p} \, dp, \qquad (28\text{-}9)$$

$$= -\int_{\varepsilon}^{p} \frac{RT}{p} \, dp + \int_{1\,\text{atm}}^{p} \frac{RT}{p} \, dp, \qquad (28\text{-}10)$$

$$= -\int_{\varepsilon}^{p} \frac{RT}{p} \, dp + RT \ln p. \qquad (28\text{-}11)$$

In Eq. 28-9 the limits of integration were interchanged, which simply reversed the sign. In Eq. 28-10 the integral from 1 atm to p was both added and subtracted.   In Eq. 28-11 the second integration was performed, and the ratio $p/(1 \text{ atm})$ was replaced by p.*   We now rewrite Eq. 28-8 in the form

$$
\begin{aligned}
\mu_i&(T, p, y_1, y_2, \ldots) \\
&= \mu_i{}^0(\text{pure } i, \text{ ideal gas, } T, 1 \text{ atm}) \\
&\quad + RT \ln y_i\, p + \int_0^p \left(v_i - \frac{RT}{p} \right) dp
\end{aligned}
\qquad \text{(general),} \quad (28\text{-}12)
$$

where we have replaced ε by 0, since the integrand is zero in the range between 0 and ε.   The result is valid for any constituent in any solution and shows how knowledge of the partial molar volume as a function of p for the temperature and concentration of interest suffices to establish μ_i.   A more compact form of Eq. 28-12 arises if it is solved for fugacity coefficient:

$$
\mu_i \equiv \mu_i{}^0 + RT \ln \varphi_i y_i\, p. \qquad (28\text{-}13)
$$

Thus

$$
RT \ln \varphi_i = \int_0^p \left(v_i - \frac{RT}{p} \right) dp \qquad \text{(general),} \qquad (28\text{-}14)
$$

which is a general expression relating φ_i to volumetric data for the mixture.

THE LEWIS FUGACITY RULE

The difficulty in using Eq. 28-12 or 28-14 to find chemical potentials is that volumetric data in the needed form,

$$
V = V(T, p, n_1, n_2, \ldots), \qquad (28\text{-}15)
$$

are rarely known for mixtures.   Such equations of state as are useful for mixtures (see Sec. 26) give p as function of T, V, and mole numbers; finding partial molar volumes from them is not easy.   An equation similar to Eq. 28-12 is needed in which pressure is presumed known.   This is Eq. 28-54, derived in Prob. 28-10.

---

* This makes sense only if p is given its value in units of atmospheres, as it is meaningless to take the log of a quantity with units.   What is the log of 1 mm of mercury?   Expressions like $\ln p$ abound in science.   However, whenever one sees anything with a log or an exponential or a trigonometric function of a quantity with units, one knows it really means something like Eq. 28-11, in which the quantity which makes the argument dimensionless has either been incorporated someplace else (as in $\mu_i{}^0$) or has been dropped.   Or else it means the writer made a mistake.

One analytic equation of state which does lend itself to use with Eq. 28-12 is Amagat's law, Eqs. 26-4 and 26-5, which says that the partial molar volume of i in the mixture is the same as the molar volume v_i' of pure i at the same temperature and total pressure:

$$v_i(T, p, y_1, y_2, \ldots) = v_i'(T, p).$$ (28-16)

If Amagat's law holds for the mixture at all pressures up to p, then φ_i is found from Eq. 28-14:

$$RT \ln \varphi_i = \int_0^p \left(v_i' - \frac{RT}{p} \right) dp,$$ (28-17)

which is exactly the same for i in the mixture and for pure i at the same T and total p. Thus

$$f_{i, \text{soln}} \equiv \varphi_{i, \text{soln}} \, y_{i, \text{soln}} \, p = y_{i, \text{soln}} \, \varphi_{i, \text{pure}} \, p,$$ (28-18)

$$f_{i, \text{soln}}(T, p, y_1, y_2, \ldots) = y_{i, \text{soln}} \, f_{i, \text{pure}}(T, p).$$ (28-19)

In Eq. 28-18, the result of Eq. 28-17 was employed. In Eq. 28-19, recognition was made that φp is the fugacity of a pure component. The result, Eq. 28-19, is the **Lewis fugacity rule** (or the **Lewis and Randall rule**). It says that the fugacity coefficient of a component in a mixture is equal to the fugacity coefficient that component would have if pure at the same temperature and total pressure as the mixture. The Lewis rule requires the validity of Amagat's law *at all pressures up to p*. As noted in Sec. 26, only at very low and very high pressures is that law expected to hold for dissimilar constituents. For dissimilar constituents, then, Eqs. 28-18 and 28-19 should be expected to hold only for dilute gases or in solutions where substance i is in large excess, say, $y_i > 0.9$. In the limit as $y_i \to 1$, the molecules of i see the same environment in the mixture as in pure i, so Eq. 28-19 then becomes exact.

VAPOR PRESSURES AND HENRY'S LAW

Real solutions may have vapor pressures either larger or smaller than would be the case if the solutions were ideal. An example showing negative departures from ideality is sketched in Fig. 28-1. It is often no easy job to describe these curves as functions of concentration. However, for dilute enough solutions the curves can always be approximated by straight lines, as shown by the dotted lines in Fig. 28-1. These straight lines,

$$p_i = k_i x_i \quad (x_i \text{ small}),$$ (28-20)

are, of course, different from those of Raoult's law, and are referred to as

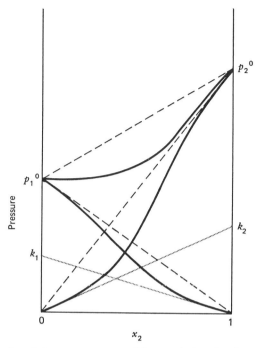

Fig. 28-1. Total and partial vapor pressures of binary solution showing negative departures from ideal behavior (dashed), and showing Henry's law curves (dotted).

Henry's law; the k_i are called Henry's law constants. They depend on the substance being dissolved, on the solvent, and on T and total p. Not only is k_1 the slope of the dotted line chosen to fit the low x_1 data but it also is the intercept of this dotted line on the $x_1 = 1$ (thus $x_2 = 0$) axis. In addition to its utility for dilute solutions, Henry's law is much used in studying the solubility of gases in liquids as functions of the pressure of the gas. For many common systems it is accurate up to 5 or 10 atm for mole fraction of dissolved gas less than, say, 0.03, unless the two components are highly dissimilar.

Thermodynamically, the basis of Henry's law is easy to understand. For the equilibrium between gas and liquid,

$$f_i^{\text{gas}} = f_i^{\text{liq}}. \tag{28-21}$$

These fugacities are

$$f_i^{\text{gas}} = \varphi_i \, p_i^{\text{gas}}, \qquad f_i^{\text{liq}} = \gamma_i \, x_i \, f_i^{\,0}. \tag{28-22}$$

If these are equated, one obtains

$$p_i^{\text{gas}} = \frac{\gamma_i \, f_i^{\,0}}{\varphi_i} \, x_i, \tag{28-23}$$

which is exact. Comparison of this with Eq. 28-20 shows that Henry's law is a statement that γ_i/φ_i be independent of concentration (and total pressure). If the gas is sufficiently dilute, $\varphi_i = 1$. However, in the liquid, one expects γ_i to be concentration independent only if the concentration of i is so small that i molecules interact only with solvent molecules and not with each other. Otherwise, as i is added to the solution, the i molecules already there interact on average with more i and less solvent. This change in molecular environment changes the activity coefficient. If i molecules " like " other i molecules better than solvent molecules, γ_i decreases with increasing x_i; if i molecules prefer solvent molecules, then γ_i increases with x_i.

One way to view Henry's law is from the vantage point of a somewhat more general expression for the activity coefficient. An expansion of the form

$$RT \ln \gamma_1 = a_1 x_2{}^2 + b_1 x_2{}^3 + c_1 x_2{}^4 + \cdots \qquad (28\text{-}24)$$

seems to be capable of fitting most experimental data for binary solutions. For many simple liquid mixtures (those whose molecules have similar size, shape, and chemical properties), it is sufficient to retain only the first term and put $a_1 = a_2 = A$:

$$\ln \gamma_1 = \frac{A}{RT} x_2{}^2, \qquad \ln \gamma_2 = \frac{A}{RT} x_1{}^2. \qquad (28\text{-}25)$$

For proof that a_1 *must* equal $a_2 (= A)$ in this approximation, see Prob. 28-9. These are called the **two-suffix Margules equations**, and the parameter A is often only weakly temperature dependent. For mixtures where A is positive, solvent and solute molecules dislike each other, compared to molecules of their own kind. Where A is negative, solvent and solute molecules might be thought of as forming something resembling a compound with each other.

If Henry's law is to hold, the concentration dependence of Eq. 28-25 (or Eq. 28-24) must be negligible. If the magnitude of A is large, this means Henry's law is valid for only the most dilute solutions, while if A is small, it is valid over a more extended concentration range.

At high pressure, one must find the pressure dependence of the Henry's law constant in order to use Eq. 28-20.

We note here that the most convenient choice of *standard state* for dilute solutions of a component whose chemical potential is written as

$$\mu_i = \mu_i{}^0 + RT \ln \gamma_i x_i \qquad (28\text{-}26)$$

may well be i whose properties are those *in the solution at infinite dilution of i*, rather than pure i. With this choice of standard state, ideal behavior would follow Henry's law, the dotted lines of Fig. 28-1. Of course, no solution could be ideal in the sense of Henry's law over the whole range $0 \leqslant x_i \leqslant 1$

(unless it was also ideal in the sense of Raoult's law), since at $x_i = 1$ the partial pressure must be p_i^0, rather than k_i. One usually picks whichever standard state lodges most of the gross concentration dependence of μ_i in the x_i of Eq. 28-26, and the least concentration dependence in the γ_i.

GIBBS-DUHEM INTEGRATIONS

Let us start with the Gibbs-Duhem equation, Eq. 13-32,

$$S\,dT - V\,dp + \sum_i n_i\,d\mu_i = 0 \tag{28-27}$$

and impose constant T and p,

$$\sum_i x_i\,d\mu_i = 0 \qquad (T \text{ and } p \text{ constant}). \tag{28-28}$$

Here we have divided through by the total number of moles to obtain x_i's rather than n_i's. Now, at constant T and p we can find $d\mu_i$ simply by differentiating Eq. 28-26:

$$d\mu_i = \frac{RT}{\gamma_i x_i}(\gamma_i\,dx_i + x_i\,d\gamma_i) \qquad (T \text{ and } p \text{ constant}). \tag{28-29}$$

This may be inserted into Eq. 28-28 to yield

$$\boxed{\sum_i x_i\,d\ln\gamma_i = 0} \qquad (T \text{ and } p \text{ constant}). \tag{28-30}$$

Advantage was taken of the fact that

$$\sum_i x_i = 1 \qquad \text{and} \qquad \sum_i dx_i = 0 \tag{28-31}$$

in obtaining Eq. 28-30. The result, Eq. 28-30, shows how the concentration dependences of the activity coefficients are interrelated.

This form of the Gibbs-Duhem equation has a number of applications. For example, for binary solutions, Eq. 28-30 is

$$x_1\,d\ln\gamma_1 + x_2\,d\ln\gamma_2 = 0. \tag{28-32}$$

If we divide through by x_1 and integrate between two concentrations we get

$$\ln\gamma_1(x_1'') = \ln\gamma_1(x_1') = -\int_{x_2'=1-x_1'}^{x_2''=1-x_1''} \frac{x_2}{1-x_2}\,d\ln\gamma_2. \tag{28-33}$$

Now suppose we know the values of γ_2 at several values of x_2 and want to know $\gamma_1(x_1)$. If we plot $x_2/(1-x_2) = x_2/x_1$ against $\ln\gamma_2$, the area under the curve between the points corresponding to two compositions is the difference in the values of $\ln \gamma_1$. If x'_2 is zero, then $\ln \gamma_1(x'_1 = 1)$ is zero and Eq. 28-33 is

$$\ln \gamma_1(x_1) = - \int_{x_2=0}^{x_2=1-x_1} \frac{x_2}{1-x_2} \, d \ln \gamma_2 . \qquad (28\text{-}34)$$

If an analytic approximation to $\gamma_2(x_2)$ is made (such as Eq. 28-25 or Eq. 28-24), the integration may be performed analytically, usually replacing $d \ln \gamma_2$ by $(d \ln \gamma_2/dx_2)dx_2$. A thorough discussion of the integration of Eq. 28-32 is given by G. N. Lewis et al.*

A useful modification of Eq. 28-32 is obtained by "division by dx_1 and recognition of the constancy of T and p" (see Eqs. A-41 to A-44):

$$x_1 \left(\frac{\partial \ln \gamma_1}{\partial x_1} \right)_{T,p} + x_2 \left(\frac{\partial \ln \gamma_2}{\partial x_1} \right)_{T,p} = 0. \qquad (28\text{-}35)$$

Since $x_2 = 1 - x_1$,

$$\left(\frac{\partial \ln \gamma_2}{\partial x_1} \right)_{T,p} = \left(\frac{\partial \ln \gamma_2}{\partial x_2} \right)_{T,p} \frac{dx_2}{dx_1} = - \left(\frac{\partial \ln \gamma_2}{\partial x_2} \right)_{T,p}, \qquad (28\text{-}36)$$

and Eq. 28-33 becomes

$$\boxed{ x_1 \left(\frac{\partial \ln \gamma_1}{\partial x_1} \right)_{T,p} = x_2 \left(\frac{\partial \ln \gamma_2}{\partial x_2} \right)_{T,p} . } \qquad (28\text{-}37)$$

If one constituent of a binary mixture obeys Henry's law, then, as noted in the discussion of Eq. 28-23, its fugacity coefficient is independent of concentration. Equation 28-35 then asserts that the fugacity coefficient of the other constituent must be concentration independent as well. Thus *in a concentration range where one component of a binary solution is ideal in the sense of Henry's law, then the other component must be also.* A corollary of this is that in the range of concentrations where the solvent is nearly pure and obeys Raoult's law the solute must obey Henry's law. Another corollary is that, if one component of a binary solution obeys Raoult's law over the entire range of $0 \leqslant x_i \leqslant 1$, then so must the other component. This is because at $x_i = 1$ the vapor pressure of i must be $p_i{}^0$; thus if it obeys Henry's law from $0 \leqslant x_i \leqslant 1$, its Henry's law constant must be $p_i{}^0$ and thus it obeys Raoult's law.

* G. N. Lewis, M. Randall, K. S. Pitzer, and L. Brewer, *Thermodynamics*, 2d ed., McGraw-Hill Book Co., New York, 1961, pp. 260–267.

One further use of Eqs. 28-34 and 28-37 is as a check for thermodynamic consistency of data.   If experimental data are available for both components of a binary solution, then these equations need not be used to find activity coefficients.   Instead, they may be used to check the data for thermodynamic consistency, the degree of agreement with Eq. 28-34 or 28-37 being a measure of reliability of the experimental results.   An example is given in Prob. 28-8.

AZEOTROPES

Sometimes the departures from ideality are so great that there is a minimum or maximum in the vapor pressure or boiling point curve.   A solution the composition of which is that of the extremum is called an *azeotropic mixture* or an *azeotrope*.   A maximum in the vapor pressure curve implies a minimum in the boiling point diagram, and vice versa.

Figure 28-2 shows a boiling point diagram, analogous to Fig. 27-4, for

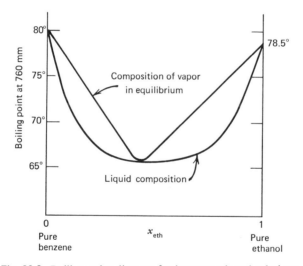

Fig. 28-2. Boiling point diagram for benzene-ethanol solutions.

solutions of benzene and ethanol.   Pure ethanol boils at 78.5°C; so if benzene-ethanol solutions were ideal, the curve of Fig. 28-2 would resemble Fig. 27-4, only it would be very flat and slightly downward sloping.   Instead there is a minimum in the boiling point at about 66°C and $x_{eth} = 0.45$.   It is easy to analyze vapor pressure and boiling point diagrams in the presence of azeotropes if one always imagines a vertical line drawn through the azeotropic concentration.   This line divides the diagram into two parts, each of which may be analyzed separately.   The azeotrope may for all practical purposes be viewed as a compound in the analysis, because, as with a compound, the vapor and the condensed phase in equilibrium with it both have the same composition. Azeotropes are not compounds, however, because they do not have definite stoichiometry. The composition at the maximum or minimum of a vapor pressure curve changes with temperature; the composition at the extremum of a boiling point diagram changes with pressure.

In the benzene-ethanol case, as example, the left-hand side of the diagram is viewed as solutions consisting of benzene and azeotrope. Since the azeotrope boils at a lower temperature than pure benzene, azeotrope is more volatile than benzene and the vapor is richer in azeotrope than the liquid is. Thus in the part of the diagram to the left of the azeotrope, the vapor line lies to the right of the liquid line. However, the right-hand side of the diagram is viewed as a mixture of azeotrope and pure ethanol. The azeotrope is more volatile than ethanol also, so in the part of the diagram to the right of the azeotrope, the vapor line lies to the left of the liquid line. Whenever an azeotrope is present, this division of the diagram into separately analyzed parts will always permit a simple analysis by using the fact that the vapor is always richer in the more volatile constituent than the liquid. However, one must view the constituents *not* as the two pure materials forming the solutions, but instead as one of the pure materials (whichever is in excess) and the azeotrope.

It is clear that if an azeotrope is formed, either maximum- or minimum-boiling, between two components, then their separation by simple fractional distillation is impossible. For example, mixtures of water (BP $= 100°C$) and ethyl alcohol (BP $= 78.3°C$) form a minimum-boiling azeotrope (BP $= 78.17°C$) of 96 weight percent alcohol. Thus solutions with alcohol weight fraction between 0 and 0.96 may be fractionally distilled to yield pure water and "pure" azeotrope, and solutions with alcohol weight fraction between 0.96 and 1.0 may be fractionated to yield pure ethanol and "pure" azeotrope. But since boiling the azeotrope at fixed pressure does not change its composition, there is no simple distillation method for going from the relatively dilute ethanol solutions formed by fermentation to alcohol of greater purity than 96 weight percent.

PARTIAL MISCIBILITY

For solutions showing positive deviations from ideality, constituent molecules prefer to be next to other molecules like themselves, rather than to those of different species. As the temperature is lowered, this effect is enhanced until finally a temperature is often reached below which over certain parts of the concentration range the mixture exists as *two phases* in

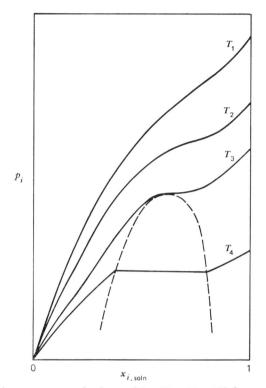

Fig. 28-3. Partial vapor pressure isotherms ($T_1 > T_2 > T_3 > T_4$) for a component above and below the critical mixing temperature ($T_c = T_3$). Dotted line shows two-phase region.

equilibrium with each other. This is sketched in Fig. 28-3. At the highest temperature, T_1, positive deviation from ideality (compare with Fig. 27-1) is evident. At the lower temperature T_2, the deviation is more evident. Finally, at T_3, the **critical mixing temperature** (or **consolute temperature**), the isotherm passes through an inflection point, completely analogous to the critical point of single-component fluids (see Fig. 18-2). At temperatures below this, there is a range of overall concentrations, shown by the dashed line of Fig. 28-3, for which two stable phases exist. The concentrations of these phases are represented by the ends of the horizontal tie line through the dashed region for the particular temperature.

The fugacity of component i must be the same in each condensed phase and in the equilibrium vapor. Sometimes the components are so unalike that the ends of the tie lines in Fig. 28-3 are nearly at $x_i = 0$ and $x_i = 1$. In that case, the two constituents are almost completely insoluble in each other at the temperature of interest. To a good approximation, then, the partial pressure in the vapor of each component is the same as the vapor pressure of that substance pure, and the total vapor pressure of the two-phase mixture is the sum of the independent pressures:

$$p_1 \approx p_1{}^0 \qquad p_2 \approx p_2{}^0 \qquad \qquad (28\text{-}38)$$

$$\text{(mutually insoluble phases)}$$

$$p \approx p_1{}^0 + p_2{}^0 \qquad \qquad (28\text{-}39)$$

The mixture will boil when the total vapor pressure equals the applied pressure, and each constituent will be present in the vapor in a mole fraction which is proportional to its pressure fraction, p_i/p. This fact is used in steam distillation to purify a compound which would not be stable (that is, might decompose, explode, or turn to tar) when heated to its normal boiling point.

EXCESS FUNCTIONS

In Eqs. 27-31, 27-39, 27-42, and 27-28, we have expressions for v_i, h_i, s_i, and μ_i as functions of T, p, and concentration in ideal solutions. It is often convenient to define so-called **partial excess functions** $v_i{}^E$, $h_i{}^E$, $s_i{}^E$, and $\mu_i{}^E$ for real mixtures as the difference between the real partial molar quantity and its value in the ideal solution with the same T, p, and concentration. For example,

$$v_i{}^E(T, p, \text{conc}) \equiv v_i(T_i, p, \text{conc}) - v_{i,\text{ideal}}(T, p, \text{conc}) \qquad (28\text{-}40)$$

and similar equations define all other partial excess functions. Comparable extensive excess functions are defined in the usual way; for example,

$$V^E(T, p, n_1, n_2, \ldots) \equiv \sum_i n_j v_j{}^E(T, p, \text{conc}). \qquad (28\text{-}41)$$

The functional relationships among the various excess functions are the same as those among the ordinary thermodynamic functions.

Perhaps the most important excess function is the partial excess Gibbs free energy $\mu_i{}^E$, which is simply

$$\mu_i{}^E \equiv \mu_i{}^0 + RT \ln \gamma_i x_i - (\mu_i{}^0 + RT \ln x_i), \qquad (28\text{-}42)$$

$$\boxed{\mu_i{}^E = RT \ln \gamma_i.} \qquad (28\text{-}43)$$

One often sees **excess Gibbs free energy** used in place of the logarithm of the activity coefficient.

The temperature and pressure derivatives of $\mu_i{}^E$ will be the same functions of the excess quantities as the derivatives of μ_i are of the regular quantities. Thus the Gibbs-Helmholtz equation, Eq. 27-38,

$$\left[\frac{\partial(\mu_i/T)}{\partial T}\right]_{p,\,conc} = -\frac{h_i}{T^2}, \qquad (28\text{-}44)$$

yields

$$\left(\frac{\partial \ln \gamma_i}{\partial T}\right)_{p,\,conc} = -\frac{h_i^E}{RT^2}. \qquad (28\text{-}45)$$

The fact that $\partial \mu_i/\partial T = -s_i$ yields

$$\ln \gamma_i + T\left(\frac{\partial \ln \gamma_i}{\partial T}\right)_{p,\,conc} = -\frac{s_i^E}{R}, \qquad (28\text{-}46)$$

and the fact that $\partial \mu_i/\partial p = v_i$ yields

$$\left(\frac{\partial \ln \gamma_i}{\partial p}\right)_{T,\,conc} = \frac{v_i^E}{RT}. \qquad (28\text{-}47)$$

The results, Eqs. 28-43 and 28-45 through 28-47, might be different depending on choices of standard state. One must use care always in analyzing μ_i^0 terms, especially their pressure dependences, in light of the pressure specified in the standard state.

HEAT OF MIXING

Since, as shown in Eq. 27-40, the heat of mixing of an ideal solution is zero, the heat of mixing of a real solution is the excess enthalpy of that solution:

$$H^E(T, p, n_1, n_2, \ldots) = \sum_j n_j[h_j(T, p, conc) - h_j^0(T, p)]. \qquad (28\text{-}48)$$

The heat absorbed per mole of solute dissolved at constant T and p is the so-called *heat of solution*, that is, it is Δh for the dissolving process. The heat of solution varies with the concentration of the solution. Suppose a solute is progressively dissolved in an originally pure solvent. As the solution composition changes, the heat of solution per mole changes.

There are several conventional ways of reporting heat of solution data. One way is to give the so-called **integral** (or total) **heat of solution**, which is defined as the enthalpy change for the process of dissolving 1 mole of pure solute in enough pure solvent to bring the resulting solution to the desired concentration. The temperature and pressure are fixed, and the integral heat of solution is a function of T, p, and concentration. It is ΔH for the following process:

$$1 \text{ mole solute} + x \text{ mole solvent} = \text{solution of mole fraction } \frac{1}{1+x}. \qquad (28\text{-}49)$$

It is the enthalpy of formation of solution of mole fraction $1/(1 + x)$ out of 1 mole of pure solute and x moles of pure solvent.

Similarly, one talks about the *integral heat of dilution*, which is Δh per mole of solute for the process of diluting a solution from one concentration to another:

1 mole solute in solution of concentration $c_1 + x$ mole solvent

$\qquad = 1$ mole solute in solution of concentration c_2.   (28-50)

Since enthalpy is a function of state, the integral heat of dilution must equal the difference between the integral heats of solution in the initial and final states.

Another approach is through the so-called *differential* (or partial) *heat of solution* which is defined as the enthalpy change per mole dissolved for the process of dissolving a small amount of pure solute in enough solution of the desired concentration that the concentration does not change appreciably in the process.   The temperature and pressure again are fixed, and the differential heat of solution also is a function of T, p, and concentration.   It is ΔH for the following process:

1 mole of solute + large amount of solution of given concentration

$\qquad =$ slightly more solution of about the same concentration.   (28-51)

It is the difference in molar enthalpy of the solute when pure and when in solution of the given concentration.

The difference between the two resides in the fact that with the differential heat of solution, the solute is dissolved in solution of the particular concentration.   However, with the integral heat of solution the solute is dissolved, starting in pure solvent which becomes more and more concentrated and which finally reaches the desired concentration.   The differential heat is commonly found from measurements of the integral heat as a function of composition.   The enthalpy change for the formation of solution containing n moles of solute in a fixed amount of solvent is $n \Delta H$, where ΔH is the integral heat of solution at that concentration.   If dn more moles of solute are added to the solution, the enthalpy change is $d(n \Delta H)$, and the ratio of this to dn is the differential heat of solution:

$$\text{Differential heat of solution} = \frac{d(n \, \Delta H)}{dn}.   \text{(28-52)}$$

Therefore if $n \Delta H$ is plotted against n, the slope of the curve at any point is the value of the differential heat of solution at that concentration.

It is noteworthy that as the temperature is increased, most fluid solutions approach ideal behavior.   This is because the attractive intermolecular potential wells become less important when thermal energies are large compared to the depths of the wells. Only the hard cores of the molecules play significant roles at high temperatures. It follows, therefore, that if the departure from Raoult's law in a binary solution is negative (that is, $\gamma_i < 1$), increasing T will increase γ_i. Such a solution will have negative excess enthalpy, as shown by Eq. 28-45. Thus mixing constituents with negative departures from ideality leads to negative absorption of heat, that is, the process is **exothermic**. This makes sense in light of the molecular picture showing such A-B solutions to have A-B pairs more stable than A-A or B-B. When the A and B molecules fall into the deeper A-B potential wells, the energy given off must leave the solution as heat, or else the temperature of the solution will increase.

An identical analysis holds for solutions showing positive departures from ideality: the mixing process is **endothermic** (that is, $Q > 0$) at constant T and p. Here, the A-B potential wells are not so deep as the A-A or B-B, so when A and B are forced up next to each other in the solution, the extra energy required must come from the surroundings in the form of heat.

PROBLEMS

28-1. The vapor pressure of pure A is 800 mm at 25°C. For pure B it is 400 mm. For a solution of A in B, the Henry law constant for A is 1200 mm; for B in A it is 600 mm. No azeotrope is formed. Sketch a reasonable form for the following plots: (a) Plot showing vapor pressure of A and of B as function of x_A. (b) Plot showing total vapor pressure of solution as function of x_A. Indicate with a dashed line on this plot approximately where the vapor composition curve must lie. (c) If $x_B = 0.03$, what is p_B?

28-2. An ideal gas mixture obeys Eq. 26-3. Calculate the partial molar volume v_i in such mixture and prove, using Eq. 28-12, that $f_i = y_i p$.

28-3. At 25°C, $K = 1.25 \times 10^6$ torrs is the Henry law constant for CO_2 dissolved in H_2O. What is the molarity of a saturated solution at one atmosphere CO_2 pressure (1 atm = 760 torrs)?

28-4. This problem is based on Fig. 28-2. A solution of benzene and ethanol of $x_{eth} = 0.20$ is heated in a pot at 1 atm of pressure. At what temperature does it boil? What is the composition of the vapor in equilibrium with the liquid? If that vapor is cooled, at what temperature does it condense to liquid? If this were part of a long fractional distillation column, what would come off the top of the column and at what temperature? What would happen to the concentration of the pot material as the distillation progressed? Ultimately, what would be separated from what? Then, suppose that instead of the solution's originally being $x_{eth} = 0.20$, that it was $x_{eth} = 0.80$, and answer the same questions over again.

28-5. The boiling point of the immiscible liquid system naphthalene-water is 98°C under a pressure of 733 mm. The vapor pressure of water at 98°C is 707 mm. Calculate the weight percent of napthalene in the distillate.

28-6. Air dissolved in water at 25°C is found to contain $\frac{1}{3}$ oxygen by volume. This means the Henry law constant for O_2 is how many times that for N_2?

28-7. A and B form a constant-boiling solution at a maximum temperature at

$x_A = \frac{1}{3}$. When a mixture of 1 mole of A and one of B is fractionally distilled, how much pure A is obtained? The boiling point of pure A is $110°C$, of pure B is $150°C$. Does the A come off the top of the column or collect in the pot?

28-8. Activity coefficients as functions of composition are available from experiment for both components in a binary mixture. The data could be checked for thermodynamic consistency by a *slope method*, that is, by comparison with Eq. 28-37. However, the slopes can rarely be measured with sufficient accuracy to make this test very meaningful. An *integral test* is easy and quantitative, though in principle less stringent. Suppose one plots $\log(\gamma_1/\gamma_2)$ versus x_1 for $0 \leqslant x_1 \leqslant 1$. Prove that the total area under the curve should be zero if pure liquids at the temperature of the mixture are chosen as the standard states:

$$\int_0^1 \ln \frac{\gamma_1}{\gamma_2}\, dx_1 = 0. \tag{28-53}$$

28-9. Prove that a_1 must equal a_2 in the approximation of Eq. 28-25.

28-10. Derive the following, which is similar to Eq. 28-14, but is useful when $p = p(V, T, n_i)$ is known, rather than $V = V(p, T, n_i)$:

$$RT \ln \varphi_i = \int_V^\infty \left[\left(\frac{\partial p}{\partial n_i} \right)_{T,V,n_{j \neq i}} - \frac{RT}{V} \right] dV - RT \ln z, \tag{28-54}$$

where $z = pV/nRT$ is the compressibility factor of the mixture, and n is the total number of moles of all species present in the mixture.

28-11. A liquid mixture contains substance A which boils at $300°K$ and substance B which boils at $500°K$. No azeotrope is formed between A and B. At $200°K$, is the vapor richer or poorer in B than the liquid?

28-12. How many grams of water would be required to steam distill 1 mole of bromobenzene? The vapor pressure of bromobenzene is 125 mm at $95°C$ and that of water is 635 mm.

28-13. At $60°C$, pure substance A has a vapor pressure of 380 mm and pure substance B of 140 mm. The mole fraction $x_B = 0.15$ is in the range of validity of Henry's law. The partial pressure of B at $x_B = 0.1$ is 20 mm. (a) What is p_B at $x_B = 0.15$? (b) What would p_B be at $x_B = 0.15$ if the solution were ideal? (c) What is γ_B at $x_B = 0.15$, using pure liquid B as the standard state for B? (d) What is γ_B at $x_B = 0.15$, using the hypothetical standard state of $x_B = 1$ but properties of an infinitely dilute solution of B in A?

28-14. The CO_2 pressure in a 12-oz soft drink is reputed to be 3.2 atm at $25°C$. How many quarts of pure CO_2 gas at 1 atm of pressure are in a soft drink? The Henry law constant for CO_2 in H_2O at $25°C$ is 1.25×10^6 torrs (1 pint = $\frac{1}{2}$ qt = 16 oz; also 2.2 lb = 1 kg).

28-15. A so-called regular binary solution has zero excess entropy and excess enthalpy given by nwx_1x_2, where w is independent of temperature. What is the activity coefficient for a component of a regular solution?

28-16. Given $\ln \gamma_2 = ax_1^2 + bx_1^3$ for a binary solution. Find $\ln \gamma_1$ as a function of a, b, and x_2.

28-17. For a mixture suppose

$$z = \frac{pV}{nRT} = \frac{pv}{RT} = 1 + \frac{B(T)}{v} + \frac{C(T)}{v^2}, \tag{28-55}$$

where the concentration dependences of B and C are given by Eqs. 26-10 and 26-11. Prove

$$\ln \varphi_i = \frac{2}{v} \sum_j y_j B_{ij} + \frac{3}{2v^2} \sum_{jk} y_j y_k C_{ijk} - \ln z, \qquad (28\text{-}56)$$

using Eq. 28-54. Note the resemblance between this and Prob. 25-19.

28-18. It is said that "Gibbs-Duhem integrations can be performed with any partial molar quantities not just μ's." Justify this statement for partial molar volumes by proving $\sum_i x_i \, dv_i = 0$ (T and p constant), which is completely analogous to Eq. 28-28.

28-19. Given the following values of ΔH:

$$HCl(g) + aq = HCl(aq), \qquad\qquad \Delta H = -17.88 \text{ kcal},$$
$$HCl(g) + 400 \text{ H}_2O(l) = HCl(400 \text{ H}_2O), \quad \Delta H = -17.70 \text{ kcal},$$

where aq means an essentially infinite amount of water. Calculate the integral heat of dilution from $HCl(50\text{H}_2O)$ to $HCl(400\text{H}_2O)$ and the integral heat of infinite dilution from $HCl(50\text{H}_2O)$. Given also the integral heat of solution of HCl gas is -17.40 kcal for $HCl(50\text{H}_2O)$.

28-20. A number of thermodynamics texts state that Eq. 28-35 implies that deviations from ideality (in the sense of Raoult's law) in binary solutions must be in the same sense (positive or negative) for both components. Prove that this is true (that is, $\ln \gamma_2$ cannot change sign) only if the slope of a plot of $\ln \gamma_2$ against x_2 never changes its sign. Show a possible plot of $\ln \gamma_2$ and a corresponding plot of $\ln \gamma_1$ against x_2 in which $\ln \gamma_2$ might change its sign, even though $\ln \gamma_1$ maintains the same sign.

29 MULTICOMPONENT EQUILIBRIA

BOILING POINT ELEVATION

This section gives thermodynamic analyses of five different multicomponent phenomena, all really varieties of phase equilibrium problems. The first is to consider how the boiling point of a solution is changed by the addition of a nonvolatile solute. A solution's **boiling point** is the temperature at which its vapor pressure equals the total applied pressure, as shown in Fig. 29-1a. The addition of a nonvolatile solute 2 to the solution lowers the fugacities of the other solution constituents and thus lowers the total vapor pressure. Since solute 2 is nonvolatile, it does not contribute to the vapor pressure. After solute 2 has been added, the only way to bring the vapor pressure back up to equal the applied pressure is to increase the temperature, as shown in Fig. 29-1b. Thus the addition of the nonvolatile

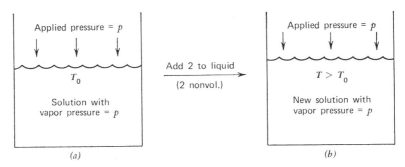

Fig. 29-1. (a) Solution at its boiling point, since its vapor pressure equals the applied pressure. (b) New higher boiling point reached after addition of nonvolatile solute 2.

solute to a solution raises the solution's boiling point.

The quantitative treatment starts from the equality of the chemical potentials in liquid and vapor phases for one of the volatile components in the solvent (it would be the solvent itself if the solvent had only one component), designated as substance 1:

$$\mu^l_{1,0} = \mu^v_{1,0} \qquad \text{(before addition of solute 2),} \qquad (29\text{-}1)$$

$$\mu^l_{1,\Delta} = \mu^v_{1,\Delta} \qquad \text{(after addition of solute 2),} \qquad (29\text{-}2)$$

$$\mu^l_{1,\Delta} - \mu^l_{1,0} = \mu^v_{1,\Delta} - \mu^v_{1,0} \qquad \text{or} \qquad \Delta\mu_1{}^l = \Delta\mu_1{}^v. \qquad (29\text{-}3)$$

The quantity $\Delta\mu_1{}^l$ may be found by integrating

$$d\mu_1{}^l = -s_1{}^l\,dT + v_1{}^l\,dp + \left(\frac{\partial\mu_1{}^l}{\partial x_2}\right)_{T,p} dx_2, \qquad (29\text{-}4)$$

from the initial boiling temperature to the final boiling temperature at the constant applied pressure along any path for which the integrand is known. If we choose the path $T_0,\ x_2 = 0 \to T,\ x_2 = 0 \to T,\ x_2$, we get

$$\Delta\mu_1{}^l = -\int_{T_0,\,x_2=0}^{T,\,x_2=0} s_1{}^l\,dT' + \int_{T,\,x_2=0}^{T,\,x_2} \left(\frac{\partial\mu_1{}^l}{\partial x_2}\right)_{T,p} dx_2 \qquad \text{(p constant),}$$

$$(29\text{-}5)$$

$$\Delta\mu_1{}^l = -\int_{T_0,\,x_2=0}^{T,\,x_2=0} s_1{}^l\,dT' + \mu_1{}^l(T, x_2) - \mu_1{}^l(T, x_2 = 0) \qquad \text{(p constant).}$$

$$(29\text{-}6)$$

The dummy temperature of integration has been written T' to prevent possible confusion. The change in μ_1 in the vapor is found by integrating

$$d\mu_1{}^v = -s_1{}^v \, dT + v_1{}^v \, dp \qquad \text{(no solute 2 in vapor)}, \qquad (29\text{-}7)$$

from T_0 to T at constant pressure:

$$\Delta\mu_1{}^v = -\int_{T_0}^{T} s_1{}^v \, dT' \qquad (p \text{ constant}). \qquad (29\text{-}8)$$

Equating Eqs. 29-6 and 29-8 in accordance with Eq. 29-3 yields

$$-\int_{T_0}^{T} [s_1{}^v - s_1{}^l(x_2 = 0)] \, dT' = \mu_1{}^l(T, x_2) - \mu_1{}^l(T, x_2 = 0) \qquad (p \text{ constant}).$$
$$(29\text{-}9)$$

The quantity in the bracket is just the molar entropy of vaporization $\Delta s_{1,\,\text{vap}}$ of pure substance 1, and the right-hand side of Eq. 29-9 is $RT \ln a_1{}^l$, simply from the definition of the activity (Eq. 27-6):

$$-\int_{T_0}^{T} \Delta s_{1,\,\text{vap}} \, dT' = -\int_{T_0}^{T} \frac{\Delta h_{1,\,\text{vap}}}{T'} \, dT'$$

$$= RT \ln a_1{}^l(T, x_2) \qquad (p \text{ constant}). \qquad (29\text{-}10)$$

This is the exact equation which relates the variables T_0, T, and x_2 in the boiling point elevation for the addition of nonvolatile solute 2 to a condensed phase.

In many cases, the temperature dependence of either $\Delta s_{1,\,\text{vap}}$ or of $\Delta h_{1,\,\text{vap}}$ can be neglected over the small temperature change of the boiling point elevation. Also the solvent in dilute solutions commonly shows ideal behavior in the sense of Raoult's law; thus a_1 is replaced by $x_1 = 1 - x_2$:

$$-\Delta s_{1,\,\text{vap}} \, \Delta T_{\text{BP}} \approx \Delta h_{1,\,\text{vap}} \frac{\Delta T_{\text{BP}}}{T} \approx RT \ln(1 - x_2)$$

$$\text{(small } \Delta T, \text{ ideal solvent behavior)}. \qquad (29\text{-}11)$$

For dilute solutions, $x_2 \ll 1$, and we can use the Taylor series expansion* of

---

* The Taylor series expansion of a function f of $1 - \varepsilon$, where $\varepsilon < 1$ is given by

$$f(1 - \varepsilon) = f(x)\Big|_{x=1} - \varepsilon \left.\frac{df(x)}{dx}\right|_{x=1} + \frac{\varepsilon^2}{2!} \left.\frac{d^2 f(x)}{dx^2}\right|_{x=1} - \cdots.$$

If the function is the logarithm, this becomes

$$\ln(1 - \varepsilon) = \ln x \Big|_{x=1} - \frac{\varepsilon}{x}\Big|_{x=1} - \frac{\varepsilon^2}{2x^2}\Big|_{x=1} + \cdots,$$

or

$$\ln(1 - \varepsilon) = -\varepsilon - \frac{\varepsilon^2}{2} - \cdots.$$

$\ln (1 - x_2)$ in powers of x_2, retaining only the lowest-order term:

$$\ln (1 - x_2) \approx -x_2 \qquad (x_2 \ll 1). \qquad (29\text{-}12)$$

Therefore Eq. 29-11 becomes simply

$$\Delta T_{\text{BP}} \approx \frac{RTx_2}{\Delta s_{1,\,\text{vap}}} \approx \frac{RT^2 x_2}{\Delta h_{1,\,\text{vap}}} \qquad (x_2 \ll 1, \Delta T \text{ small, ideal solvent behavior}).$$

$$(29\text{-}13)$$

FREEZING POINT DEPRESSION

The *freezing point* (FP) of a liquid is the temperature at which it coexists with a solid under the applied total pressure, as shown in Fig. 29-2a. This

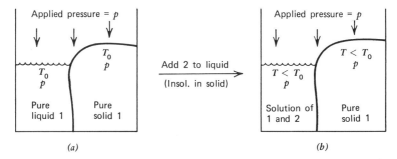

Fig. 29-2. (a) Pure 1 at its freezing point under pressure p.   (b) Solution of 1 and 2 at new depressed freezing point reached after addition of solute 2 (insoluble in solid).

means the chemical potential of each component which can enter both solid and liquid phases will have the same value in both phases.   It commonly happens that, when a solution begins to freeze, the solid that is formed is almost pure solvent; very few solute molecules enter the crystal lattice. In other words the solute is nearly insoluble in the solid phase.   The addition of solute 2 to the liquid (usually) lowers the fugacity of the solvent (substance 1); it does not change the solid phases, however, since it does not enter the solid.   The only way, then, to maintain the equality of chemical potential of substance 1 in the two phases is to decrease the temperature, as shown in Fig. 29-2b.   Thus the addition of solute to the solution lowers the liquid's freezing point.

The quantitative treatment is analogous to that for boiling point elevation. Equations 29-1, 29-2, and 29-3 govern the equilibria before and after the addition of substance 2.   The change in μ_1 in the liquid due to the addition of substance 2 and to the depression of the freezing point is given by Eq. 29-6. Since the analog to nonvolatility is here the insolubility of substance 2 in the solid, the phase analogous to the vapor of Eqs. 29-7 and 29-8 is in this case the solid. Thus the equation analogous to Eq. 29-9 is

$$- \int_{T_0}^{T} [s_1{}^s - s_1{}^l(x_2 = 0)] \, dT' = \mu_1{}^l(T, x_2) - \mu_1{}^l(T, x_2 = 0) \qquad (p \text{ constant}).$$

$$(29\text{-}14)$$

The bracketed quantity is the negative of the molar entropy of fusion, $-\Delta s_{1,\text{fus}}$, of pure substance 1:

$$\int_{T_0}^{T} \Delta s_{1,\text{fus}} \, dT' = \int_{T_0}^{T} \frac{\Delta h_{1,\text{fus}}}{T'} \, dT' = RT \ln a_1{}^l(T, x_2) \qquad (p \text{ constant}).$$

$$(29\text{-}15)$$

This is the exact equation, analogous to Eq. 29-10, which relates T_0, T, and x_2 in the freezing point depression phenomenon for the addition to solvent of a solute 2, which is insoluble in the solid phase which freezes out.

Here we can use the same approximations as in Eqs. 29-11 to 29-13 to yield

$$\Delta T_{\text{FP}} \approx - \frac{RTx_2}{\Delta s_{1,\text{fus}}} \approx - \frac{RT^2 x_2}{\Delta h_{1,\text{fus}}} \qquad (x_2 \ll 1, \Delta T \text{ small, ideal solvent behavior}).$$

$$(29\text{-}16)$$

The reason there is a difference in sign between Eqs. 29-10 and 29-15, or between Eqs. 29-13 and 29-16 is that the solution is the *low*-temperature (liquid) phase in the liquid-vapor equilibrium involved in the boiling point elevation. The solution is the *high*-temperature (liquid) phase in the solid-liquid equilibrium involved in the freezing point depression.

SOLUBILITY OF SOLIDS IN LIQUIDS

Examination of Fig. 29-2b shows that this is precisely the equilibrium condition reached when pure solid 1 is dissolved in solute 2 at pressure p and temperature T until the solution becomes saturated. Quantitatively, Eq. 29-15 gives the relationship between the activity of substance 1 in the saturated liquid to the heat of fusion of pure solid 1. The temperature T_0 in Eq. 29-15 is the melting point of pure solid 1 under the pressure p. We can rewrite Eq. 29-15, replacing a_1 by $\gamma_1 x_1$, solving for x_1, and noting that $T < T_0$:

$$x_1 = \frac{1}{\gamma_1} \exp \left[- \int_{T}^{T_0} \frac{\Delta h_{1,\text{fus}}(T') \, dT'}{RTT'} \right] \qquad (p \text{ constant}). \qquad (29\text{-}17)$$

This is the exact expression for the maximum solubility of 1 in 2 at T and p.

Neglecting the effect of nonideality, γ_1, on this result, we can note certain qualitative consequences of the structure of Eq. 29-17:

1. For a given choice of solvent and solute, increasing the temperature increases the solubility. At T_0, substance 1 is infinitely soluble in solute 2, since a "solution" of $x_1 = 1$ can coexist with solid 1.

2. Given the solvent and two different solids with similar melting points, the solid with the lower heat of fusion has the higher solubility.

3. Given the solvent and two solids with similar heats of fusion, the lower melting of the two has the higher solubility.

Although it is needed in Eq. 29-17, one usually does not know the enthalpy of fusion as a function of T, but instead one knows it at only a single value of T, say, at T_0. Unless the melting point of the solute is more than, say, $100°C$ above T, one can simply forget the T dependence of the heat of fusion, with the result

$$x_1 = \frac{1}{\gamma_1}\left(\frac{T_0}{T}\right)^{-\Delta h_{1,\,fus}(T_0)/RT} \qquad (p \text{ constant, } \Delta h \text{ independent of } T). \qquad (29\text{-}18)$$

If the heat of fusion must be known more precisely, it may be corrected with heat capacity data by using Eq. 25-34.

A plot of temperature against maximum solutility x_1 will rise until at $x_1 = 1$ the freezing point of pure solid 1 is reached. This is shown in the right-hand part of Fig. 29-3. The region labeled I is filled with horizontal

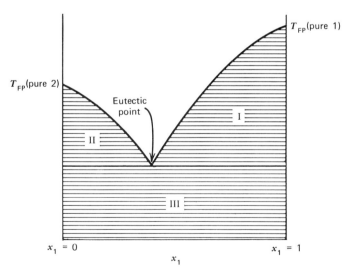

Fig. 29-3. Freezing point diagram for typical binary mixtures.   Pressure is fixed.

tie lines. A system whose overall T and x_1 place it within region I will exist in two phases represented by the end points of the tie line through the system point. One phase is pure solid 1, the other phase is melt of the appropriate composition. It is clear, however, that exactly the same analysis must hold for solutions of solid 2 in substance 1. This region of high concentration of substance 2 is shown in the left-hand part of Fig. 29-3. Here again, region II is filled with horizontal tie lines, and systems whose representative points lie within region II must exist as two phases, pure solid 2 and the saturated solution of appropriate concentration. At the unique temperature (for given p) where these solubility curves meet, mixtures of pure solid 1, pure solid 2, and liquid solution can coexist. This is called the **eutectic point** and a solution of that composition is called the **eutectic mixture**. At temperatures below the eutectic, in region III, pure solid 1 coexists with pure solid 2; the tie lines run from $x_1 = 0$ to $x_1 = 1$.

Significant effects of nonideality in Eqs. 29-17 or 29-18 should be expected whenever the component molecules differ in nature or size. In nonpolar solutions, the γ's are generally greater than unity and the mutual solubilities are less than would otherwise be expected. In solutions where polar or chemical attractions exist between the component molecules, the γ's may be less than unity, so the solubilities would be enhanced.*

OSMOTIC PRESSURE

There exist so-called **semipermeable membranes**, both in nature and man made, which have the property of letting some kinds of molecules through but not others. Thus if equilibrium is established across such a membrane, the chemical potential of any substance the passage of which is permitted must be the same on both sides of the membrane. If it were not, that kind of material would diffuse from the side with higher chemical potential across to the side with the lower. Obviously no such restriction is present for the constituents for which the membrane is impermeable.

Suppose we have pure substance 1 on one side of a membrane which is permeable only to 1. This is shown in Fig. 29-4a. On the other side of the membrane is a solution with 2 dissolved in 1. If the temperature and pressure are the same on the two sides, 1 will flow through the membrane from the pure side into the solution because the 1 on the pure side is more concentrated, thus has a higher chemical potential, than it does in the

---

* An extensive discussion including experimental data is given by J. M. Prausnitz, *Molecular Thermodynamics of Fluid-Phase Equilibria*, Prentice Hall, Englewood Cliffs, N.J., 1969, Chapter 9.

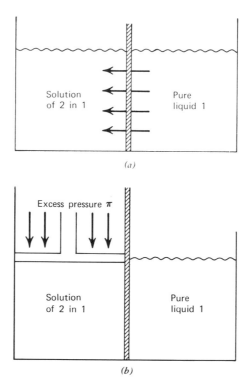

Fig. 29-4. (*a*) Osmotic transport of 1 across semipermeable membrane. (*b*) Osmotic transport stopped by application of excess pressure π on the solution.

solution.   Of course, if something were done to the solution to raise μ_1 in it, for example, increasing its pressure, this imbalance could be ended.   The amount by which the pressure on the solution must be increased over the pressure on the pure liquid in order just to stop the flow of 1 through the membrane and achieve equilibrium is called the **osmotic pressure** of the solution.

Calculation of osmotic pressure is straightforward.   Nothing has happened to the pure liquid on the right hand side of the membrane during the process of adding 2 to the left side and then increasing the pressure on the left side.   Therefore its chemical potential is unchanged:

$$\Delta\mu_1^{\text{pure}} = 0 \qquad (T \text{ constant}). \tag{29-19}$$

In the solution we have dissolved substance 2 and changed the pressure:

$$d\mu_1^{\text{soln}} = v_1^{\text{soln}} \, dp^{\text{soln}} + \left(\frac{\partial\mu_1^{\text{soln}}}{\partial x_2}\right)_{p,\,T} dx_2 \qquad (T \text{ constant}). \tag{29-20}$$

This may be integrated along any path for which the integrand is known, for example, $x_2 = 0$, $\pi = 0 \rightarrow x_2$, $\pi = 0 \rightarrow x_2$, π:

$$\Delta\mu_1^{soln} = \int_{0,\,x_2}^{\pi,\,x_2} v_1^{soln}\,dp^{soln} + \mu_1^{soln}(x_2,\,\pi = 0) - \mu_1^{soln}(x_2 = 0,\,\pi = 0)$$
$$(T\text{ constant}).\quad(29\text{-}21)$$

Equating Eqs. 29-21 and 29-19 yields

$$-\int_{0,\,x_2}^{\pi,\,x_2} v_1^{soln}\,dp^{soln} = RT\ln a_1^{soln}(x_2,\,\pi = 0)\quad(T\text{ constant}),\quad(29\text{-}22)$$

which is the general relationship among π, x_2, and T for osmotic equilibria.

If the liquid is sufficiently incompressible for the molar volume of 1 in the solution to be considered constant, the integral is performed immediately:

$$-v_1^{soln}(x_2)\pi = RT\ln a_1^{soln}(x_2,\,\pi = 0)$$
$$(T\text{ constant, incompressible solution}).\quad(29\text{-}23)$$

If the solution is not incompressible, v_1 may be corrected by using the isothermal compressibility. Since in dilute solutions the solvent commonly shows ideal behavior, Eq. 29-23 is approximated by

$$v_1^{soln}(x_2)\pi = -RT\ln(1 - x_2)$$
$$(T\text{ constant, incompressible solution, ideal solvent behavior}).\quad(29\text{-}24)$$

For dilute solutions, Eq. 29-12 may be used to obtain

$$v_1^{soln}(x_2)\pi = x_2\,RT\quad(T\text{ constant; incompressible, ideal, dilute solution}).$$
$$(29\text{-}25)$$

As a convenient mnemonic, Eq. 29-25 may be multiplied through by the total number of moles n in the solution. For dilute solutions, n times the molar volume of solvent is approximately the total solution volume V. Also nx_2 is n_2, the number of moles of substance 2 dissolved in the solution. The result is

$$\pi V = n_2\,RT\quad(T\text{ constant; incompressible, ideal, dilute solution}),$$
$$(29\text{-}26)$$

which is convenient to remember because of its similarity to the ideal-gas law.

A few special cases of the action of osmotic flow warrant comment. If one simply has solution on one side of a membrane permeable only to solvent and has solvent on the other, as long as no pressure buildup occurs on the solution side, solvent will flow through forever. Since no amount of diluting of the solution can ever make it have $x_2 = 0$, the solvent flow would in principle never stop.

If the dimensions of the chamber holding the solution were fixed so that as solvent flowed in the liquid level moved up, something analogous to Fig. 29-5 would be observed.   Once the hydrostatic head of pressure of the column of solution of height h just equals π for that diluted solution,

$$\pi = \frac{n_2 RT}{V} = \rho^{\text{soln}} gh, \tag{29-27}$$

equilibrium is reached and the flow of 1 through the membrane ceases.   In Eq. 29-27, ρ is the mass of solution M^{soln} per unit volume V, so the height h is given simply by

$$M^{\text{soln}} gh = n_2 RT. \tag{29-28}$$

Many people interested in biological processes find it convenient to refer to the "osmotic pressure" as a *property* of a solution relative to pure solvent, much like its vapor pressure.   In this usage it is not the excess pressure applied to a solution to make μ of its solvent equal to μ for the pure solvent.   Instead, it is simply the excess pressure *that would have to be applied* to make these μ's equal.   Thus osmotic pressure is simply another measure of the concentration of a solution, indirectly a measure of the activity of the solvent (activity in the sense of $a = \exp[(\mu - \mu^0)/RT]$, although this is certainly correlated to activity in the sense of being "active"), just like the freezing point depression or the boiling point elevation.   This dual usage of "osmotic pressure" has led some scientists to serious misconceptions of the osmotic phenomenon.

Suppose the chamber holding the solution was a red blood cell, which is filled with various salts and other dissolved molecules.   If this cell is put into a solution whose water solvent has greater activity than the water in the cell, then water will flow into the cell through its wall.   With the word

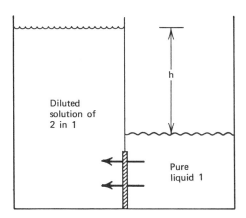

Fig. 29-5. Osmotic transport continuing until the hydrostatic pressure head equals π.

usage mentioned previously, the "osmotic pressure inside the cell (phase β) exceeds the osmotic pressure outside it " (phase α) by an amount $\Delta\pi = \pi^\beta - \pi^\alpha$. The value of $\Delta\pi$ is obtained by writing Eq. 29-22 for phase α and again for phase β, with $\pi^\beta = \pi^\alpha + \Delta\pi$. With incompressible solutions, the integrals may be done immediately to yield

$$v_1{}^\beta \, \Delta\pi = RT \ln \frac{a_1{}^\alpha}{a_1{}^\beta} + (v_1{}^\alpha - v_1{}^\beta)\pi^\alpha \qquad (T \text{ constant, incompressible solution}).$$

$$(29\text{-}29)$$

At equilibrium, the pressure inside the cell must be greater than that outside by an amount $\Delta\pi$.   The cell expands until the excess pressure inside it due to the elasticity of its walls reaches 6.5 cm of H_2O and its volume is 1.6 times the normal value.   Red blood cell walls can stand no greater stress than that, so the cell will burst whenever $\Delta\pi$ is greater than 6.5 cm H_2O.   Since molar volumes of condensed phases are small, huge osmotic pressures arise from relatively small concentrations of solutes.   For example, if n_2/V is 1 mole/l, π is 24.6 atm.   When one realizes that 1 atm is 1034 cm of H_2O, one must be impressed by the extraordinarily strict requirements (see Prob. 29-9) on any living thing regarding the concentration of dissolved salts in its body fluids so that osmosis will not lead to fatal results.

SOLUBILITY OF LIQUIDS AND SOLIDS IN GASES

We now consider how the partial vapor pressure of a constituent in a condensed solution is changed when the total applied pressure is increased. The increase is commonly made by simply raising the partial pressure of some constituent not of direct interest in the problem.   We are studying then the solubility of components of solid or liquid solutions in gases as a function of the total gas pressure.   We find that at high pressures the enhancement of vapor pressure is extremely large, at low pressures unimportant.

We start with a condensed phase in equilibrium with vapor at temperature T and pressure p_0, as shown in Fig. 29-6a.   The equalities of chemical potentials of substance 1 before and after the addition of gas 2 are written

$$\mu_{1,0}^v = \mu_{1,0}^c, \qquad \mu_{1,\Delta}^v = \mu_{1,\Delta}^c, \qquad (29\text{-}30)$$

or

$$\mu_{1,\Delta}^v - \mu_{1,0}^v = \mu_{1,\Delta}^c - \mu_{1,0}^c. \qquad (29\text{-}31)$$

The left-hand side of Eq. 29-31 is $RT \ln f_{1,\Delta}^v / f_{1,0}^v$, since the same choice of standard state is made for substance 1 in the vapor before and after the addition of gas 2.   The right-hand side is the change in μ_1 in the condensed phase

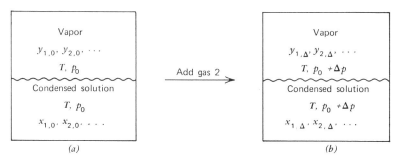

Fig. 29-6. (a) Multicomponent phase equilibrium, similar to Figure 25-2b. (b) New equilibrium reached after the addition of gas 2 to the system of (a).

caused by the increase in total pressure from p_0 to $p_0 + \Delta p$ and by possible change in concentration caused by the solution of gas 2 in the condensed phase. We have already found this change in the derivation of Eq. 29-21, so Eq. 29-31 can be written

$$RT \ln \frac{f_{1,\Delta}^v}{f_{1,0}^v} = \int_{p_0}^{p_0 + \Delta p} v_1^c(x_{1,\Delta}) \, dp + RT \ln \frac{a_{1,\Delta}^c(p_0)}{a_{1,0}^c(p_0)}. \qquad (29\text{-}32)$$

We may solve this for the ratio of partial vapor pressures, using $f_1^v = \varphi_1^v p_1$:

$$\ln \frac{p_{1,\Delta}}{p_{1,0}} = \ln \frac{\gamma_{1,\Delta}(p_0)x_{1,\Delta}\varphi_{1,0}^v}{\gamma_{1,0}(p_0)x_{1,0}\varphi_{1,\Delta}^v} + \int_{p_0}^{p_0 + \Delta p} \frac{v_1^c(x_{1,\Delta})}{RT} \, dp. \qquad (29\text{-}33)$$

This is the result desired, and it exactly relates the new and old partial vapor pressures of substance 1 under the new and old total pressures, p_0 and $p_0 + \Delta p$.

The solubility of gases in solid phases is usually negligible; in liquids, matters are not always so simple and the gas may dilute the pure liquid 1. Where gas 2 solubility in the condensed phase can be neglected, Eq. 29-33 simplifies:

$$\ln \frac{p_{1,\Delta}}{p_{1,0}} = \ln \frac{\varphi_{1,0}^v}{\varphi_{1,\Delta}^v} + \int_{p_0}^{p_0 + \Delta p} \frac{v_1^c(x_{1,\Delta})}{RT} \, dp \qquad \begin{array}{l}\text{(gas 2 insoluble in} \\ \text{condensed phase).}\end{array} \qquad (29\text{-}34)$$

A rewriting of Eq. 29-34 is common for the case where the condensed phase is pure liquid 1, where p_1^s is the ordinary saturation vapor pressure of 1 at the temperature T, and where the final pressure $p^s + \Delta p$ is simply p:

$$p_1 \equiv y_1 p \equiv p_1^s E, \qquad (29\text{-}35)$$

where E is called the **enhancement factor**,

$$E = \frac{\varphi_1^s}{\varphi_1} \exp\left(\int_{p_1^s}^{p} \frac{v_1^c}{RT} \, dp\right) \qquad (T \text{ constant}). \qquad (29\text{-}36)$$

The exponential part of E is called the **Poynting correction**; it is always positive, but rarely larger than 2 or 3. The fugacity coefficient $\varphi_1{}^s$ in the saturated vapor is usually near unity, since $p_1{}^s$ is usually small. The dominant term is φ_1, the fugacity coefficient of substance 1 in the high-pressure gas mixture. This is almost always less than unity, and sometimes leads to enhancement factors of 10^3 or more. In fact, values of E up to 10^{12} have been observed for the solution of solid oxygen in hydrogen, almost entirely due to the small value of φ_1! Here is a phenomenon for which the ideal-gas approximation is almost always poor. One needs at least the second virial contributions to φ_1 for even quite low pressures (see Prob. 25-19 and 28-17), and greater precision is needed for relatively high pressures.*

PROBLEMS

29-1. Sometimes boiling point elevation problems are worked with the so-called *molal boiling point constant* K_b, defined by the equation $\Delta T_{\mathrm{BP}} = K_b m$, where m is the molality of the solution. Calculate the molal boiling point constant for water which has a boiling point of 100.00°C, a heat of vaporization of 539.7 cal/g, and a molecular weight of 18.02 g/mole.

29-2. Calculate the freezing point constant defined by $\Delta T_{\mathrm{FP}} = -K_f m$ for water as in Prob. 29-1. The heat of fusion at the normal freezing point is 79.7 cal/g.

29-3. What is the osmotic pressure of a sucrose solution in water, presuming an ideal solution and calculated from the knowledge that at this temperature, 30°C, the solution vapor pressure is 31.207 torrs? Pure water at 30°C has vapor pressure of 31.824 torrs. Is this at all consistent with the solution's being one molal?

29-4. One mole of liquid mercury (200.6 g) is placed in an evacuated 20-l flask at the normal boiling point of mercury (630.0°K). At this temperature the vapor pressure of the mercury (1 atm) is too small for all the liquid to evaporate in the fixed volume. However, all the liquid mercury can be made to evaporate by pumping inert gas into the vessel, thus raising the vapor pressure of the mercury. Calculate the pressure of inert gas required assuming that all gases and gas mixtures are ideal (an unreasonable assumption) and that liquid mercury is incompressible and has a density of 12.8 g/cm³.

29-5. Calculate the solubility of naphthalene in benzene at 25°C, assuming an ideal solution is formed. The melting point of naphthalene is 80°C, and its heat of fusion is 4610 cal/mole.

29-6. What fractional change $\delta p/p_0$ in vapor pressure will be found for water at 25°C if 1 atm of excess air pressure is applied to the water? Neglect the solubility of air in the water; treat gases as ideal.

29-7. Calculate the molal fractional vapor pressure lowering and the molal osmotic pressure increase for ideal solutions in water. Using these results plus those of Probs. 29-1 and 29-2, rank these four methods of molecular weight determination in decreasing order of expected usefulness.

29-8. A solution of 0.1 g of compound AB in 1 l of H_2O at 0°C has an osmotic pressure of 0.02 atm. The compound is known to be completely dissociated into

* J. M. Prausnitz, *op. cit.*, section 5.14.

A and B in this solution. What is the approximate molecular weight of compound AB?

29-9. The freezing point depression of human blood serum is 0.535°C. What is the maximum percent increase in this value that could occur in the blood before the red cells would burst? See the discussion of Eq. 29-29.

29-10. Benzene boils at 80.1°C at 1 atm. What is the boiling point elevation if 1 mole percent of a nonvolatile impurity is present? Assume an ideal solution and use Trouton's rule.

29-11. At the eutectic point for the binary Bi-Cd system how many phases are in equilibrium? How many degrees of freedom are there? What is the physical meaning of this result?

29-12. Find the molar heat of fusion of benzene (C_6H_6) if 0.2 mole of analine lowers the freezing point of 1000 g of benzene from -69°C by 0.99°C.

29-13. What is the solubility of solid naphthalene in benzene at 25°C, assuming an ideal solution? Naphthalene melts at 80°C; its heat of fusion is 4610 cal/mole.

29-14. The heat of fusion of substance A is 6000 cal/mole. Substance B dissolves in A but does not ionize in A. Calculate the change in freezing point on adding 0.01 mole of B to a mixture of 10 moles of B and 30 moles of A at 25°C. Assume A and B form an ideal solution and that B is nonvolatile.

29-15. What is the exact expression for the osmotic pressure which must be applied to an incompressible regular solution of x_2 mole fraction of substance 2? The equilibrium is with pure substance 1 through a membrane permeable only to substance 1.

29-16. The following are reliable data on sea water: freezing point, -1.91°C; boiling point, 100.56°C; osmotic pressure, 23.1 atm; vapor pressure lowering 15.2 mm. What is meant by the last value given? Are these values mutually consistent? Why?

29-17. What is the minimum amount of work needed to desalinate 1000 l of sea water? At 10 cents/kwhr, what would be the minimum cost of producing 1000 l of pure water from the sea? How does this result compare with the cost of distillation if none of the heat of vaporization is reclaimed in heat exchangers? The heat of vaporization of water is about 540 cal/g. See Prob. 29-16 for data on sea water and Prob. 29-18 for a hint.

29-18. Since Fig. 29-4b represents an equilibrium reached by imposing the pressure π on the solution, increasing that pressure further should lead to a flow of substance 1 from the solution into the pure liquid 1. This is sometimes called *negative osmosis*; the membrane acts like a molecular sieve. It is an attractive approach to water desalination, reversible and thus of maximum efficiency. Suppose the process is performed on sea water and the brine concentrated to four times its concentration in original sea water. How will the cost per 1000 l compare with that of Prob. 29-17, where the salt concentration in the solution remained essentially constant? Assume all activity coefficients remain the same as the concentration is increased. If the brine were concentrated to eight times its original concentration, how would the cost compare?

29-19. A tree has its roots in a pool of pure water at 25°C at elevation zero. What is the maximum height to which the tree could transport water to be evaporated from the leaves, as a function of the relative humidity r? Presume that the sole driving force for the work against gravity, Mgh, is the difference in chemical potential of the water in the pool and in the air. If the relative humidity is 90%, what is this maximum height?

30   CHEMICAL EQUILIBRIUM

THE CRITERION FOR CHEMICAL EQUILIBRIUM

Thermodynamics permits the prediction of the equilibrium conditions that will be encountered for various chemical reactions by using knowledge of only a relatively few properties of the reactants and products. This powerful application is of utmost importance to chemists, biologists, geologists, and many engineers. In this section, we limit our treatment to a few basic features of chemical thermodynamics; to do otherwise would make the book far too long. However, it is valuable for students to become familiar with other books having more extensive treatments, references to which are given herein.

The main difference between a chemical reaction and a phase equilibrium of the type already studied is the balanced, stoichiometric chemical equation which relates changes in the amounts of the chemical species present. For example, during the reaction between hydrogen and bromine to give hydrogen bromide, the change in the number of moles of HBr in the system is always the negative of twice the change in the mole numbers of H_2 and Br_2. This might be written $H_2 + Br_2 = 2HBr$, in which case we would say that the **stoichiometric number** for HBr, ν_{HBr}, is $+2$, for H_2 it is -1, and for Br_2 it is -1. Stoichiometric numbers are conventionally chosen to be positive for products of the reaction as written and negative for reactants. If the reaction were written $HBr = \frac{1}{2}H_2 + \frac{1}{2}Br_2$, the stoichiometric numbers would be $\nu_{HBr} = -1$, $\nu_{H_2} = \frac{1}{2}$, $\nu_{Br_2} = \frac{1}{2}$, but exactly the same chemical reaction would be described by the two balanced equations. For a given balanced chemical equation, the stoichiometric number for compound i is written ν_i.

In Sec. 15 the principle that an isolated system evolves to an equilibrium condition characterized by maximum entropy was exploited to find the criteria for various kinds of equilibria. Here we do the same thing for chemical equilibria. In an isolated system the change in entropy

$$dS = \frac{1}{T} dU + \frac{p}{T} dV - \frac{1}{T} \sum_i \mu_i \, dn_i \qquad (30\text{-}1)$$

becomes simply

$$dS = -\frac{1}{T}\sum_i \mu_i \, dn_i \qquad (U \text{ and } V \text{ constant}). \qquad (30\text{-}2)$$

If this change can be made positive by having reactants form products or by having products form reactants, that will happen. The n_i will change through the reaction in such a way that S is maximized. Finally, at equilibrium, the chemical potentials of the various species in the reaction vessel will be such that dS in Eq. 30-2 is zero, because S itself is a maximum:

$$dS = -\frac{1}{T}\sum_i \mu_i \, dn_i = 0 \qquad (U \text{ and } V \text{ constant, equilibrium}). \qquad (30\text{-}3)$$

If each mole number could change independently of the others, then the dn_i would be arbitrary and Eq. 30-3 would imply that $\mu_i = 0$ for all i. This is precisely what occurs in the case of electromagnetic radiation contained in a cavity at equilibrium. There is no conservation of photons; they are continually being created or destroyed at the walls or by any dust particle or other particle in the system. Thus radiation at equilibrium has $\mu_i = 0$ for each wavelength. However, in chemical reactions, the stoichiometry couples the n_i. If $n_i{}^0$ is the number of moles of i originally put into the vessel, at any later time the number n_i still there is given by

$$n_i = n_i{}^0 + \nu_i \xi. \qquad (30\text{-}4)$$

The ξ is a variable representing the extent of the reaction; sometimes it is called the **progress variable**. It tells how many times the stoichiometric numbers of moles of reactants have formed into the stoichiometric numbers of moles of products, that is, how many times the reaction as written has progressed. If ξ is positive, the reaction has gone from left to right as written; if negative, it has gone from right to left. The magnitude of ξ can be small ($\ll 1$) or large ($\gg 1$). A change in ξ by $+1$ means that the mole number of each species has changed by its stoichiometric number, products being formed, reactants used up. Conversely, if $\Delta\xi = -1$, the number of moles of each species present has changed by the negative of its stoichiometric number, and "products" have been used up to form "reactants;" the reaction has proceeded from right to left as written.

An example might illustrate this. For the reaction $H_2 + Br_2 = 2HBr$, suppose one started by mixing 10 moles of each of the three compounds together. If the reaction proceeded from left to right, after $\Delta\xi = 1$, there would be $n_{H_2} = n_{Br_2} = 9$ and $n_{HBr} = 12$ remaining; after $\Delta\xi = 2$, there would be $n_{H_2} = n_{Br_2} = 8$ and $n_{HBr} = 14$; after $\Delta\xi = 5$, there would be $n_{H_2} = n_{Br_2} = 5$ and $n_{HBr} = 20$; etc. If the reaction proceeded from right to left, after $\Delta\xi = -2\frac{1}{2}$, there would be $n_{H_2} = n_{Br_2} = 12\frac{1}{2}$ and $n_{HBr} = 5$, etc. Thus for chemical reactions the n_i are not separately variable quantities. Instead, once the material is added to the system, ξ is the independent variable. Changes in mole numbers dn_i can be related to $d\xi$ by differentiating Eq. 30-4:

$$dn_i = v_i \, d\xi \quad \text{(closed, chemically reacting system)}. \quad (30\text{-}5)$$

Then the criterion for equilibrium, Eq. 30-3, becomes

$$dS = -\frac{1}{T} \sum_i \mu_i v_i \, d\xi = 0 \quad (U \text{ and } V \text{ constant, chemical equilibrium}).$$

$$(30\text{-}6)$$

The only way in which this can be zero for all $d\xi$ is if

$$\boxed{\sum_i v_i \mu_i = 0} \quad \text{(criterion for chemical equilibrium)}. \quad (30\text{-}7)$$

Equation 30-7 is the basic equation of chemical equilibrium. It is a relationship among the intensive quantities μ_i which must hold at each point in the system. Although it was derived for an isolated system, it must hold even if T and p, rather than U and V, are fixed, since the μ's are functions of T, p, and composition, and the final equilibrium condition must not depend on how the equilibrium was reached.

Suppose one has a mixture of H_2, Br_2, and HBr in a reaction vessel. A chemical reaction will proceed, either H_2 with Br_2 to form HBr or HBr dissociating to form H_2 and Br_2, until finally, at the existing T and p, the concentrations become such that

$$2\mu_{HBr} - \mu_{H_2} - \mu_{Br_2} = 0. \quad (30\text{-}8)$$

The reaction then stops. We note that this condition for equilibrium is the same regardless of how the stoichiometric equation happens to have been written; for example, $\frac{1}{2}\mu_{H_2} + \frac{1}{2}\mu_{Br_2} - \mu_{HBr} = 0$ is exactly the same relationship among the μ's as $2\mu_{HBr} - \mu_{H_2} - \mu_{Br_2} = 0$. Clearly, this must be the case, since the reactants and products do not care how scientists choose to write the balanced chemical reaction!

Since μ_i is the partial molar Gibbs free energy of substance i at a particular T, p, and composition, $2\mu_{HBr} - \mu_{H_2} - \mu_{Br_2}$ is the change in Gibbs free energy for the process of forming two moles of HBr and using up one mole each of H_2 and Br_2, all at the T, p, and composition of the reaction vessel. We note that so extensive a reaction could not occur in a finite-sized vessel without the various concentrations being changed. An actual process which accomplished the previous formation would have to be carried out in a huge reaction vessel containing reactants and products at exactly the same concentrations, temperature, and pressure as the reaction vessel, but with reactants and products in essentially infinite quantities so that the composition would be almost unchanged by adding or subtracting a few moles.   For brevity, one often writes

$$\sum_i v_i \mu_i \equiv \Delta G \qquad (T, p, \text{ and composition fixed}), \qquad (30\text{-}9)$$

that is, ΔG is the change in Gibbs free energy for forming the stoichiometric numbers of moles of products from the stroichiometric numbers of moles of reactants at the fixed T, p, and composition found in the system.   Another way to view ΔG is to note (see Eq. 13-29) that

$$G = \sum_i n_i \mu_i = \sum_i (n_i^0 + v_i \xi)\mu_i, \qquad (30\text{-}10)$$

where we have used Eq. 30-4.   Differentiation of Eq. 30-10 with respect to ξ at constant T, p, and composition yields

$$\left(\frac{\partial G}{\partial \xi}\right)_{T, p, \text{comp}} = \sum_i v_i \mu_i(T, p, \text{comp}) \equiv \Delta G. \qquad (30\text{-}11)$$

So ΔG is the derivative of the Gibbs free energy for the system with respect to the progress variable.

At equilibrium,

$$\Delta G = 0 \qquad (\text{criterion for chemical equilibrium}). \qquad (30\text{-}12)$$

Suppose ΔG is positive for a particular mixture of reactants and products. The reaction will proceed in such a way as to reduce its value.   The values of μ for the products are too high relative to those for reactants, so the reaction is driven from right to left ($\Delta \xi < 0$) by a positive ΔG.   If ΔG is negative, the reaction proceeds from left to right ($\Delta \xi > 0$).   In other words, if the system can lower its Gibbs free energy at fixed T and p by reacting from left to right as written, it will do so; if it can lower its Gibbs energy by reacting from right to left it will do so.   The equilibrium condition

$$\left(\frac{\partial G}{\partial \xi}\right)_{T, p, \text{comp}} = 0 \qquad (\text{criterion for chemical equilibrium}), \qquad (30\text{-}13)$$

arising from Eqs. 30-11 and 30-12 is simply a statement that at fixed T, p, and initial composition, the chemical reaction will proceed until a state of minimum Gibbs free energy is reached.

The negative of ΔG is often called the **chemical affinity** in the reaction vessel.

One way to measure ΔG for a reaction is to set up the system so the reaction is opposed by a voltage from an electric cell.   Electrochemical cells can often be made so that the reaction can be driven slightly in one direction by an infinitesimal increase in applied voltage and slightly in the other direction by an infinitesimal decrease in voltage. That is, the reaction can be performed reversibly through the use of the cell.

Now Eq. 30-1 was written with the presumption that the only work interaction between system and surroundings was compressional. If there is electrical work also, it was noted in Eq. 5-7 that $\mathscr{E} \, dq$ must be added to $-p \, dV$, where $\mathscr{E}$ is taken here to be the potential difference through which the charge dq is moved:

$$dS = \frac{1}{T} \, dU + \frac{p}{T} \, dV - \frac{\mathscr{E}}{T} \, dq - \frac{1}{T} \sum_i \mu_i \, dn_i. \qquad (30\text{-}14)$$

If now one follows through the same analysis that led to Eq. 30-6, one gets

$$dS = -\frac{1}{T} \sum_i \nu_i \mu_i \, d\xi - \frac{\mathscr{E}}{T} \, dq = 0 \qquad (30\text{-}15)$$

instead. The charge dq can be related to the change $d\xi$ in the progress variable as follows: Let n be the number of equivalents of charge that pass from one electrode to the other when the reaction as written is performed using the cell. A faraday is the magnitude of the charge on Avogadro's number of electrons (an equivalent of charge), and its value is

$$\mathscr{F} = 96{,}493 \text{ coul/equiv} = 23{,}062.4 \text{ cal/volt-equiv}. \qquad (30\text{-}16)$$

The n is just the number of electrons transferred during the course of the reaction as written. For example, $H_2 + Br_2 = 2HBr$ has $n = 2$, since both hydrogen atoms change from an oxidation state of 0 to $+1$ while both bromines change from 0 to -1. If the reaction was written $\frac{1}{2}H_2 + \frac{1}{2}Br_2 = HBr$, instead, n would be 1. For the electrochemical cell, then,

$$dq = n\mathscr{F} \, d\xi, \qquad (30\text{-}17)$$

and Eq. 30-15 simplifies to

$$\Delta G = -n\mathscr{F}\mathscr{E} \qquad \text{(reactants opposed by reversible cell).} \qquad (30\text{-}18)$$

Here the reactants and products of such composition as to lead to the chemical potentials in $\Delta G = \sum_i \nu_i \mu_i$ are kept from reacting by the reversible cell with an applied voltage of $\mathscr{E}$.

The voltage $\mathscr{E}$ is clearly a measure of the "desire" of a reacting mixture to react. Thus ΔG is also. Measurement of voltages in electrochemical cells is a most important way of determining the values of chemical potentials, and thus of measuring activities.

It is clear that, if the concentrations of reactants and products in the cell are those characteristic of equilibrium, then the voltage needed to prevent reaction must be zero. If it were not, the "equilibrium" chemical system could spontaneously do electrical work on the surroundings, thus violating the second law.

STANDARD STATES

The criterion for equilibrium, Eq. 30-7, takes a more intuitive form if expressed in terms of activities (Eq. 27-6):

$$\mu_i(T, p, \text{comp}) \equiv \mu_i^0(\text{standard state}) + RT \ln a_i(T, p, \text{comp, standard state}).$$

$$(30\text{-}19)$$

We have noted previously that the choice of standard state for a substance in a particular phase will be made on the practical grounds of convenience. The value of an activity is clearly meaningless unless the standard state is specified with respect to which it is calculated. It is well to summarize here the most commonly used standard states in dealing with chemical reactions:

Each component in a *gaseous* phase is assigned as its standard state the state of unit fugacity, that is, the pure gas in the hypothetical ideal condition at temperature T and pressure 1 atm. Thus $a_i \to p_i$ as $\rho_i \to 0$. Numerically, a_i and f_i are always the same, then, for gases, except that a_i is dimensionless and f_i is measured in atmospheres. Real gases at 1 atm usually are very close to having $f = 1$ atm.

For a *pure solid or liquid* or for the *solvent in a solid or liquid solution* the most convenient standard state is the pure solid or liquid solvent at temperature T and 1 atm of pressure. For such a solvent, $a_1 \to x_1$ as $x_1 \to 1$ if the pressure is 1 atm. Unless the actual pressure is extremely high, however, the p dependence of μ_i in condensed phases can be neglected because of the small molar volume in the term $v_i \, dp$. Thus we say that an ideal solution (in the sense of Raoult's law) has $a_i = x_i$ regardless of the pressure, while, in fact, $a_i = x_i$ rigorously only at 1 atm.

For a *solute in a solid or liquid solution*, convenience might dictate any one of several possible choices of standard state. If the concentration range over which the solution is being studied is a wide one, then solvent and solute are really on a par in one's thinking. Ideality in the sense of Raoult's law is a more useful behavior for comparison than ideality in the sense of Henry's law, so a standard state is chosen similar to that for the solvent: pure solid or liquid solute at temperature T and 1 atm pressure. For such a "solute," $a_2 \to x_2$ as $x_2 \to 1$ (again, rigorously, only at $p = 1$ atm, but to a good approximation, at all low to moderate pressures also).

For a *solute in a relatively dilute solution*, Henry's law is the most convenient kind of "ideal" behavior with which to make comparison. Therefore the behavior of the solute in the standard state is that at infinite dilution. This is analogous to the case for gases, where the standard state is the hypothetical state whose properties are those of infinite dilution but whose pressure is 1 atm. Similarly, for dilute solutions, the properties of the solute in its standard state are those of infinite dilution at 1 atm of pressure, but the concentration is chosen for convenience to be either (a) mole fraction of unity, (b) composition of 1 molal (mole/1000 g of solvent), or (c) composition of 1 molar,* depending on how solute concentration is being measured in practice. Clearly this is a hypothetical standard state, since the solute is unlikely still to be obeying Henry's law at 1 molar, 1 molal, or $x_2 = 1$. With choice (a), $a_2 \to x_2$ as $x_2 \to 0$. With choice (b), a_2 approaches the numerical value of the molality of 2 as $x_2 \to 0$. With choice (c), a_2 approaches the numerical value of the molarity of 2 as $x_2 \to 0$. For very dilute solutions, molarity and molality are both proportional to mole fraction (see Prob. 30-1), so choices (b) and (c) both represent adherence to Henry's law. However, at appreciable concentrations, choice (a) still retains Henry's law as ideal behavior, but choices (b) and (c) no longer do. Sometimes a solution which is so dilute that $a_i \approx c_i$ is called a **perfect solution**.

Occasionally when dealing with condensed phases, it is convenient to use one of the foregoing choices for standard state, with the pressure specified as the applied pressure p, rather than 1 atm.

THE EQUILIBRIUM CONSTANT

We now use Eq. 30-19 to represent ΔG for a chemical reaction:

$$\Delta G \equiv \sum_i \nu_i \mu_i \equiv \sum_i \nu_i \mu_i{}^0 + RT \sum_i \nu_i \ln a_i, \qquad (30\text{-}20)$$

$$\Delta G \equiv \Delta G^0 + RT \ln (a_1{}^{\nu_1} \cdot a_2{}^{\nu_2} \cdot a_3{}^{\nu_3} \cdots), \qquad (30\text{-}21)$$

$$\Delta G \equiv \Delta G^0 + RT \ln Q. \qquad (30\text{-}22)$$

In Eq. 30-20, substitution for μ_i was made using Eq. 30-19. In Eq. 30-21, the symbol ΔG^0 was introduced for $\sum_i \nu_i \mu_i{}^0$; ΔG^0 is the change in Gibbs free energy for a process starting with the stoichiometric number of moles of each reactant, separated, and in their standard states; and ending with the stoichiometric number of moles of each product, separated, and in their standard states. In Eq. 30-22, the product of activities is given the symbol Q. The resulting identity, Eq. 30-22, is sometimes called the **reaction isotherm**.

* The molality is a more useful concentration unit than molarity for solutions where T and p are variables, since the molality is independent of T and p and molarity is not.

As we noted previously, the value of ΔG in a mixture of reactants and products is a measure of the mixture's nonequilibriumness, of its desire to react; it is the driving force for the chemical reaction in much the way ΔT drives heat flow. The value of ΔG is expressed in Eq. 30-22 in terms of the activities of the reactants and products. As the reaction proceeds, the concentrations and thus these activities change until finally at equilibrium, $\Delta G = 0$, and Eq. 30-22 becomes

$$\boxed{\Delta G^0 = -RT \ln K} \qquad (T \text{ constant, equilibrium}). \qquad (30\text{-}23)$$

The value of Q at equilibrium has been written K to emphasize the constancy of

$$K \equiv (a_1{}^{v_1} \cdot a_2{}^{v_2} \cdot a_3{}^{v_3} \cdots)_{\text{eqlm}} \qquad (30\text{-}24)$$

at a given T, independent of pressure and independent of the initial concentrations of the reactants and products in the reaction vessel. However much of these substances is put into the vessel initially, the reaction always proceeds until the same value for Eq. 30-24 is reached. The K is called the **equilibrium constant** for the reaction.

While it might appear superficially that these results depend on how one chooses to write the stoichiometric equation, that is not the case. For the reaction $H_2 + Br_2 = 2HBr$, ΔG^0 is $2\mu^0_{HBr} - \mu^0_{H_2} - \mu^0_{Br_2}$ and K is the equilibrium value of $a^2_{HBr}/a_{H_2}a_{Br_2}$. If we wrote the reaction $HBr = \frac{1}{2}H_2 + \frac{1}{2}Br_2$, then ΔG^0 would be $\frac{1}{2}\mu^0_{H_2} + \frac{1}{2}\mu^0_{Br_2} - \mu_{HBr}$, which is $-\frac{1}{2}$ times the first ΔG^0. However, K would be the equilibrium value of $a^{1/2}_{H_2} a^{1/2}_{H_2}/a_{HBr}$, which is the reciprocal of the square root of the first equilibrium constant. Thus the logarithm of the second is $-\frac{1}{2}$ that of the first, so Eq. 30-23 is consistent however the reaction is written.

It is interesting to replace ΔG in Eq. 30-22 by $-n\mathscr{F}\mathscr{E}$ from Eq. 30-18, and also to replace ΔG^0 by $-n\mathscr{F}\mathscr{E}^0$. Here $\mathscr{E}^0$ is the voltage applied to an electrochemical cell which just prevents reaction between reactants and products, all separated and in their standard states. The result for Eq. 30-22 is

$$\mathscr{E} = \mathscr{E}^0 - \frac{RT}{n\mathscr{F}} \ln Q, \qquad (30\text{-}25)$$

which is the familiar **Nernst equation** (see Prob. 30-2). The value of $\mathscr{E}^0$ can be determined for a wide variety of reactions (at 25°C) from extensive tabulations* of $\mathscr{E}^0$ for various reactants and products in their standard states

---

* W. M. Latimer, *Oxidation Potentials*, 2d ed., Prentice-Hall, Englewood Cliffs, N. J., 1952. For a less extensive, but thorough, discussion of galvanic cells and electrode potentials, see G. N. Lewis, M. Randall, K. S. Pitzer, and L. Brewer, *Thermodynamics*, 2d ed., McGraw-Hill Book Co., New York, 1961, chapter 24.

reacting with a standard hydrogen electrode, which is a device for oxidizing hydrogen to H^+ ions, with both H_2 and H^+ at unit activity. If these half-cell potentials are added and subtracted appropriately to yield the stoichiometric equation, the hydrogen reference cancels out and the result is $\mathscr{E}^0$ for the desired reaction.

One need not obtain equilibrium constants from tables of half-cell potentials and

$$\mathscr{E}^0 = \frac{RT}{n\mathscr{F}} \ln K. \tag{30-26}$$

Instead, one may take advantage of extensive tables of the so-called **standard Gibbs free energy of formation** $\Delta G_f{}^0$ of compounds at various temperatures. The value of $\Delta G_f{}^0$ is ΔG for the process of forming 1 mole of the substance in its standard state at the temperature T out of the requisite numbers of gram-atoms of the pure elements in their standard states, also at T. It thus is ΔG for the reaction

$$\text{Pure elements at } T = 1 \text{ mole pure compound at } T. \tag{30-27}$$

Given a table listing $\Delta G_f{}^0(T)$ for all the reactants and products in any proposed reaction, $\Delta G^0(T)$ may be found for the reaction by adding $\Delta G_f{}^0$ for each mole of product and subtracting $\Delta G_f{}^0$ for each mole of reactant:

$$\Delta G^0(T) = \sum_i v_i \, \Delta G_f{}^0(T). \tag{30-28}$$

This is valid because chemical reactions conserve matter, and the same number of gram-atoms of each element are in the products as in the reactants.

Example 30-1. Suppose one was thinking of performing the water gas reaction $CO_2 + H_2 = H_2O + CO$ at 400°K and wanted to know the value of K. The Gibbs free energies of formation at 400°K are found from tables to be -35.01 kcal/mole for CO, -94.33 kcal/mole for CO_2, and -53.52 kcal/mole for H_2O. Calculate K for this gas reaction.

Solution: Use of Eq. 30-28 would permit our writing $\Delta G^0 = (-53.52 - 35.01 + 94.33)$ kcal/mole $= 5.80$ kcal/mole, since $\Delta G_f{}^0$ for elements (in this case hydrogen) must be zero. Another way to view this is to write down the reactions which the $\Delta G_f{}^0$'s represent and add them all up, noting that $\Delta G_f{}^0$ changes sign when the reaction is written backward:

$H_2 + \frac{1}{2}O_2 = H_2O$	$\Delta G^0 = -53.52$ kcal/mole
$C + \frac{1}{2}O_2 = CO$	$\Delta G^0 = -35.01$ kcal/mole
$CO_2 = C + O_2$	$\Delta G^0 = 94.33$ kcal/mole
$H_2 + CO_2 = H_2O + CO$	$\Delta G^0 = 5.80$ kcal/mole

Thus K can be found from $\Delta G^0 = -RT \ln K$:

$$\log K = \frac{-\Delta G^0}{2.3\,RT}$$

$$= -\frac{(5800\text{ cal})(\text{deg-mole})}{(2.3)(\text{mole})(1.987\text{ cal})(400\text{ deg})}$$

$$= -3.17 = -4 + .83,$$

$$K = 6.8 \times 10^{-4}.$$

Reactions which as written have negative ΔG^0 are said (especially by biologists) to be **exergonic**. Reactions with $\Delta G^0 > 0$ are called **endergonic**. It is frequently said that exergonic processes " are spontaneous and can go to completion," while " endergonic processes cannot go unless energy is supplied from without." The statement, however, is extremely misleading. At $400°K$, as shown above, the reaction $H_2 + CO_2 = H_2O + CO$ is endergonic by 5.8 kcal and $a_{H_2O}\,a_{CO}/a_{H_2}\,a_{CO_2}$ at equilibrium is 6.8×10^{-4}. Nevertheless, if H_2 and CO_2 were mixed at $400°K$, they would react to give H_2O and CO until Q equaled 6.8×10^{-4}. Of course, the reverse reaction in which H_2 and CO_2 form from H_2O and CO is *favored*; the equilibrium lies way over in the direction of H_2 and CO_2. But the reaction will go either way, depending on the initial concentrations. No reaction can go to *completion* so long as Q can reach the value K in the course of the reaction, because K can never be either 0 or ∞ since ΔG^0 can never be either $+\infty$ or $-\infty$. Reactions *can* go to completion, but this is usually accomplished by the removal of a product as it is formed, thereby its activity is kept abnormally low, and Q is prevented from reaching K.* This is, of course, facilitated by a large negative value of ΔG^0 and a resulting large value of K, also by increasing the concentrations of the reactants as much as possible. However, many reactions in laboratories, industries, and living organisms go essentially to completion despite positive values of ΔG^0. Despite all this, the following is a useful rule of thumb for chemists contemplating proposed reactions: If ΔG^0 is negative, the reaction is promising. If ΔG^0 is between 0 and 10 kcal, the reaction is questionable, but worth checking. If ΔG^0 is greater than 10 kcal, it is useful only under very unusual conditions. Of course, *all* reactions go in such a direction (at constant T and p) as to decrease G.

Sometimes, instead of knowing ΔG^0 directly, one computes it from values of the standard entropy and enthalpy changes, ΔS^0 and ΔH^0:

$$\Delta G^0 = \Delta H^0 - T\,\Delta S^0 \qquad (T\text{ constant}). \tag{30-29}$$

---

* This is a reason why living organisms must maintain the physical transport of materials of various kinds into and out of the regions where important reactions are occurring within them.

HEAT OF REACTION AND TEMPERATURE
DEPENDENCE OF EQUILIBRIUM CONSTANT

The heat absorbed by a system during a constant pressure process is ΔH for that process, if only p-V work is done. Thus the formation of the stoichiometric numbers of moles of products, separated and in their standard states, from the stoichiometric numbers of moles of reactants, separated and in their standard states, at constant T and p requires that Q be

$$\Delta H^0 = \sum_i v_i h_i^0 \tag{30-30}$$

for $\Delta \xi = 1$. Here h_i^0 is the molar enthalpy of i in its standard state. Of course, Eq. 30-30 is not the value of $Q = \Delta H$ for $\Delta \xi = +1$ in the reaction vessel, since reactants and products there are not separated and in their standard states. Thus h_i in the mixture is different from h_i^0.

If ΔH for the reaction is positive in the direction the reaction is proceeding, then $Q > 0$, heat is absorbed (at constant T and p) and the reaction is said to be **endothermic**. If ΔH is negative, $Q < 0$, heat is given off, and the reaction is said to be **exothermic**. These definitions are sometimes based on the sign of ΔH^0 rather than ΔH, there being little conflict between them if the reaction always is going from left to right as written.

The value of K depends on T, but not on p. Its T dependence is easily found by rewriting Eq. 30-23:

$$-R \ln K = \sum_i v_i \frac{\mu_i^0}{T}, \tag{30-31}$$

$$R\left(\frac{\partial \ln K}{\partial T}\right) = \sum_i \frac{v_i h_i^0}{T^2}. \tag{30-32}$$

In Eq. 30-32, the differentiation with respect to temperature was made and the Gibbs-Helmholtz equation, Eq. 27-38, was used to simplify the right-hand side:

$$\left(\frac{\partial \ln K}{\partial T}\right) = \frac{\Delta H^0}{RT^2} \tag{30-33}$$

An equivalent form, perhaps more convenient, is*

$$\left(\frac{\partial \ln K}{\partial (1/T)}\right) = -\frac{\Delta H^0}{R}. \tag{30-34}$$

This result is sometimes called the **Van't Hoff equation.**

---

* The relationship between $\partial f/\partial x$ and $\partial f/\partial(1/x)$ is easily found by using the chain rule:

$$\frac{\partial f}{\partial x} = \frac{\partial f}{\partial(1/x)}\frac{d(1/x)}{dx} = -\frac{1}{x^2}\frac{\partial f}{\partial(1/x)}.$$

If one plots $\ln K$ against $1/T$, as shown in Fig. 30-1, the slope is $-\Delta H^0/R$. The fact that the curves are not quite straight lines illustrates the fact that ΔH^0 for the reaction varies with temperature. If we use the following imprecise but intuitive notation for exothermic and endothermic processes,

$$\text{Reactants} = \text{products} + \text{heat} \qquad (T \text{ constant, exothermic}), \quad (30\text{-}35)$$

$$\text{Reactants} + \text{heat} = \text{products} \qquad (T \text{ constant, endothermic}), \quad (30\text{-}36)$$

we can observe that, in either case, increasing T causes the equilibrium to shift

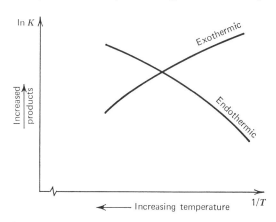

Fig. 30-1. Plots of $\ln K$ versus $1/T$ for two reactions, one exothermic and one endothermic.

in the endothermic direction, toward reactants for exothermic processes and toward products for endothermic ones. Sometimes the name of **Le Chatelier** is mentioned in connection with this conclusion.*

Integration of the Van't Hoff equation requires knowing the T dependence of ΔH^0. Sometimes only a single value of ΔH^0 is known and the approximation of its constancy with T must be made:

$$\int d(\ln K) = \int -\frac{\Delta H^0}{R} \, d\left(\frac{1}{T}\right) \qquad (30\text{-}37)$$

$$\ln K = -\frac{\Delta H^0}{RT} + c \qquad (\Delta H^0 \text{ independent of } T), \qquad (30\text{-}38)$$

or

---

* Le Chatelier's principle, roughly stated, is that equilibrium systems subjected to a stress will change their equilibrium in order in part to alleviate the stress. The principle, while sometimes intuitive and helpful, can prove misleading.

$$\ln \frac{K(T_2)}{K(T_1)} = -\frac{\Delta H^0}{R}\left(\frac{1}{T_2} - \frac{1}{T_1}\right) = \frac{\Delta H^0(T_2 - T_1)}{RT_2 T_1} \qquad (\Delta H^0 \text{ independent of } T).$$

$$(30\text{-}39)$$

In Eq. 30-38, the indefinite integral was taken, the result being the equation for the plots in Fig. 30-1 if they are approximated by straight lines. In Eq. 30-39, the definite integral was taken, the result being the relationship between any two points on the straight-line plots of $\ln K$ versus $1/T$. The resemblance of these equations to the Clausius-Clapeyron equation is not coincidence, since the vapor pressure equilibrium resembles chemical equilibrium.

Unfortunately, if the Van't Hoff integration must be made over even rather small temperature ranges, the Kirchhoff equation, Eq. 25-26, is needed to correct for the temperature dependence of ΔH^0. Over moderate temperature ranges,

$$\Delta H^0(T') = \Delta H^0(T_0) + (T' - T_0)\,\Delta C_p{}^0 \qquad (30\text{-}40)$$

suffices, but over extended ranges, Eq. 25-26 must be used with knowledge of the T dependence of the heat capacities. Sometimes tabulated values of $\Delta H^0(T)$ are available, so a numerical integration of the Van't Hoff equation is possible.

Example 30-2.* Certain large South American beetles of the genus *Brachinus* are called *bombardier beetles* because they spray a hot solution of quinone in water at any predator which is attacking them. Their unerring marksmanship in directing the hot, irritating spray in any direction and their ability to repeat the explosive discharge as often as 20 to 30 times in rapid succession protect them in a most striking manner.

In a reservoir the beetle stores an aqueous solution of 25% H_2O_2 and 10% hydroquinone and methylhydroquinone mixed. At discharge, some of this fluid is squeezed into a mixture of enzyme catalysts which causes the oxidation of the hydroquinones:

$$\text{Hydroquinone } (aq) + H_2O_2(aq) = \text{quinone } (aq) + 2H_2O(l), \qquad (A)$$

and the decomposition of the excess H_2O_2:

$$H_2O_2(aq) = H_2O(l) + \tfrac{1}{2}O_2(g). \qquad (B)$$

The standard enthalpy of reaction B is -22.6 kcal/mole, but ΔH^0 is not known for A as such. However,

$$H_2(g) + \tfrac{1}{2}O_2(g) = H_2O(l) \qquad (C)$$

* D. J. Aneshansley, T. Eisner, J. M. Widom, and B. Widom, *Science* **165** : 61 (1969).

has $\Delta H^0 = -68.3$ kcal/mole.   Also measurements of standard electrode potentials for the reaction

$$\text{Hydroquinone } (aq) = \text{quinone } (aq) + H_2(g) \qquad \text{(D)}$$

at different temperatures show $\partial(\mathscr{E}^0/T)/\partial(1/T) = -0.92$ volt for the plot of $\mathscr{E}^0/T$ against $1/T$.   Calculate the temperature of the spray and its heat content (that is, the heat absorbed by the predator in cooling the spray down to ambient temperature).

Solution: The heat content per milligram of solution is found from ΔH^0 for reaction A for the quinone formation (assume the substituted and unsubstituted hydroquinones to be identical) and reaction B for the excess H_2O_2. We have to find $\Delta H_A{}^0$ by noting that its stoichiometry is identical to the sum of reactions B, C, and D; thus its ΔH^0 must be too.   This follows from the fact that H is a function of state, independent of path, so reaction A could be performed by doing B first, then C, and finally D.   This useful property of heats of reaction, no doubt already familiar to the reader, is often called **Hess' law**.   In order to get $\Delta H_D{}^0$ from the potential data given, we must derive the appropriate relationship, using Eqs. 30-34 and 30-26:

$$\Delta H^0 = -R\frac{\partial \ln K}{\partial(1/T)} \qquad \text{and} \qquad \ln K = \frac{n\mathscr{F}\mathscr{E}^0}{RT}. \qquad (30\text{-}41)$$

Therefore

$$\Delta H^0 = -n\mathscr{F}\frac{\partial(\mathscr{E}^0/T)}{\partial(1/T)}, \qquad (30\text{-}42)$$

$$\Delta H_D{}^0 = \frac{-(2 \text{ equiv})(23{,}060 \text{ cal})(-0.92 \text{ volt})(\text{kcal})}{(\text{mole})(\text{volt-equiv})(1000 \text{ cal})} = 42.4 \text{ kcal/mole.}$$
$$(30\text{-}43)$$

Thus $\Delta H_A{}^0 = -22.6 - 68.3 + 42.4 = -48.5$ kcal/mole at 25°C.   We neglect the temperature dependence of the ΔH^0's.

One milligram of solution of 10% hydroquinone (molecular weight = 110), 25% H_2O_2 (molecular weight = 34), and 65% H_2O (molecular weight = 18) contains*

$$\frac{(1 \text{ mg solution})(0.1 \text{ mg hydroquinone})(\text{mmole/hydroquinone})}{(\text{mg/solution})(110 \text{ mg hydroquinone})}$$

$$= 9.1 \times 10^{-4} \text{ mmole hydroquinone} \qquad (30\text{-}44)$$

and

---

* The problem gives concentrations in percent, but does not specify whether the percentage is calculated by weight, by volume, or by numbers of moles.   This is a common occurrence. The best advice in such cases is to presume that weight per cent is meant.

$$\frac{(1\text{ mg solution})(0.25\text{ mg } H_2O_2)(\text{mmole } H_2O_2)}{(\text{mg solution})(34\text{ mg } H_2O_2)} = 73.6 \times 10^{-4} \text{ mmole } H_2O_2.$$

$$(30\text{-}45)$$

This much hydroquinone undergoing reaction A contributes

$$\Delta H = (9.1 \times 10^{-4} \text{ mmole})(-48.5 \text{ cal/mmole}) = -0.044 \text{ cal,} \quad (30\text{-}46)$$

and the excess H_2O_2 (64.5×10^{-4} mmole) undergoing reaction B contributes

$$\Delta H = (64.5 \times 10^{-4} \text{ mmole})(-22.6 \text{ cal/mmole}) = -0.146 \text{ cal} \quad (30\text{-}47)$$

for a total heat content of 0.19 cal/mg of solution.

Since the specific heat of water is 1 cal/deg-g, the 0.19 cal could heat the milligram of solution up to its boiling point, approximately 100°C, using up about $(75 \text{ deg})(1 \text{ cal/deg-g})(0.001 \text{ g}) = 0.075$ cal in the process, and still have 0.12 cal left over. The heat of vaporization of water is about 600 cal/g, so the fraction x of solution that could be vaporized at 100°C is found from $(x)(0.001 \text{ g})(600 \text{ cal/g}) = 0.12$ cal; thus x is about 0.2.

Experiments with bombardier beetles show these calculations to be excellent, with a heat content of about 0.22 cal/mg, and a discharge temperature of 100°C. It is interesting that in the evolution of these beetles the oxidation reaction A to quinone proved useful to them because of quinone's irritating effect on predators. However, as Eqs. 30-46 and 30-47 show, the heating effect of the quinone formation is not significant. Heating due to the breakdown of excess H_2O_2 so improved the beetle's defenses that it paid to develop separate sets of enzymes to accomplish both reactions. Furthermore, the oxygen gas produced by reaction B dramatically increases the pressure, thus permitting the explosive discharge.

Example 30-3. The *enthalpy of combustion* at 25°C of the hydrocarbon $CH_3(CH_2)_n H$ is ΔH^0 for the reaction

$$CH_3(CH_2)_n H(g) + (\tfrac{3}{2}n + 2)O_2(g) = (n + 1)CO_2(g) + (n + 2)H_2O(l).$$

The vaporization of liquid water at 25°C absorbs 10.52 kcal/mole. The value of C_p is known for CO_2, H_2O, and N_2 gases in the form of Eq. 19-29. Calculate the maximum temperature in the flame if the gaseous hydrocarbon is burned in air (1 mole O_2 to 4 moles N_2) at 25°C.

Solution: The needed approximation is that the total system whose temperature is increased in the flame is the $(n + 1)$ moles of CO_2, $(n + 2)$ moles of H_2O, and the accompanying $4(\tfrac{3}{2}n + 2)$ moles of N_2. We recognize that there must be heat lost to the surroundings, incomplete combustion, excess air mixed in the flame, and chemical dissociations at the high temperatures, all of which tend to lower the temperature from that calculated. Still, the model of an adiabatic, constant-pressure combustion will yield a maximum value for the flame temperature.

For $Q = 0$ and $p = $ constant,

$$\Delta H = 0 = \Delta H(\text{combustion}) + \int_{298^0}^{T} C_{p,\,\text{gas mixture}}(T')\, dT', \qquad (30\text{-}48)$$

which is the basic equation for calculating adiabatic flame temperatures. In terms of the data given for this problem, ΔH of combustion is ΔH^0 for the reaction written above (which will be a large negative quantity) plus $(n + 2)$ times 10.52 kcal, which changes the water into the vapor state at 25°C. The value of C_p for the gas mixture is obtained by summing the expression of the form of Eq. 19-29 for each of the gases in the system, that is, $n + 1$ moles of CO_2, $n + 2$ moles of H_2O, and $6n + 8$ moles of N_2. The result is a cubic equation in T, which may be solved by successive approximations.

Calculated flame temperatures are commonly around 2000°C for many hydrocarbons, H_2 and CO. Experimental values are often about 100° lower than calculated.

TABLES OF THERMODYNAMIC DATA

There are innumerable tabulations of thermodynamic properties of organic and inorganic compounds and mixtures scattered through a wide variety of books and journals. Unfortunately, many of these data are unreliable, usually because of faulty methods of preparation of the samples (seen only in retrospect). Other experimental difficulties and even the use of different fixed constants and conversion factors help cloud the results. The best places to begin a search for self-consistent data, so presented that the accuracy can be estimated, are the following: For organic compounds: D. R. Stull, E. F. Westrum, Jr., and G. C. Sinke, *The Chemical Thermodynamics of Organic Compounds*, John Wiley & Sons, New York, 1969. For inorganic compounds: D. R. Stull, I. Carr, J. Chao, T. E. Dergazarian, L. A. du Plessis, R. E. Jostad, S. Levine, F. L. Oetting, R. V. Petrella, H. Prophet, and G. C. Sinke, *JANAF Thermochemical Tables*, Clearinghouse for Federal Scientific and Technical Information, Springfield, Va., 1966. These books contain references to other collections, and they have useful discussions of the problems inherent in using tabulated thermodynamic data. A large number of tables of various kinds of thermodynamic data, especially handy for quick reference, are given by G. N. Lewis, M. Randall, K. S. Pitzer, and L. Brewer, *Thermodynamics*, 2d ed., McGraw-Hill Book Co., New York, 1961; see Subject Index under "Tables."

We now examine certain functions conventionally tabulated and the rationale back of their choice.   Suppose one wanted $\Delta H^0(T)$ for a reaction and had a tabulation of $h_i^0(T) - h_i^0(0°K)$ for reactants and products, where Eq. 24-8 was used to find the values.   From such a table one could find only $\Delta H^0(T) - \Delta H^0(0°K)$ and not the desired $\Delta H^0(T)$.   Therefore, for convenience with chemical reactions, *the pure constituent elements that make up the compound are used as reference, rather than the compound itself.*   The **standard enthalpy of formation** $\Delta H_f^0(T)$ of compound i is the enthalpy change for the process of making one mole of i in its standard state at T out of the pure separated elements in their stable states at T and 1 atm of pressure.   Since matter is conserved in chemical reactions, the reference enthalpies of the elements always cancel out, and we can write

$$\Delta H^0(T) = \sum_i v_i \, \Delta H_{f,\,i}^0(T), \qquad (30\text{-}49)$$

in analogy to Eq. 30-30.

As it happens, the enthalpy of reaction at a given temperature to form compound i out of its elements is less accurately known (since *dissociation energies* are hard to measure precisely) than $h_i^0(T) - h_i^0(0°K)$ from Eq. 24-8. Therefore, one often finds tabulations of either $h_i^0(T) - h_i^0(T^0)$ or $[h_i^0(T) - h_i^0(T^0)]/T$, as well as $\Delta H_{f,\,i}^0(T)$.   Here T^0 is a reference temperature chosen for convenience to be either $0°K$ or $298.15°K$.   The reason $h_i^0(T^0)$ is subtracted out is so this part of $h_i^0(T)$ will remain unchanged as better experiments yield better values for $h_i^0(T^0)$.   The reason for dividing by T is simply to make the quantity vary more slowly as a function of T and thus permit easier interpolation when T's are given perhaps only every 100°.   The quantity $[h_i^0(T) - h_i^0(T^0)]/T$ is sometimes called the **enthalpy function** (abbreviated ef).

Also one sometimes finds reference to the so-called **absolute enthalpy** of a compound, defined to be ΔH for the process of going from the pure crystalline elements which make up 1 mole of the compound at $0°K$ and ending with the compound in the state under consideration.   The only difference between absolute enthalpies and enthalpies of formation is that the constituent elements in the former start at $0°K$, while in the latter they start at T.

Things are simpler, in a way, for entropies.   The **absolute entropy** of a compound is defined to be ΔS for the process of going from the pure crystalline elements which make up the compound at $0°K$ and ending with the compound in the state under consideration.   This is completely analogous to the absolute enthalpy discussed above, with Eq. 21-28 replacing Eq. 24-8, except for one simplifying feature.   In contrast to the case of enthalpies, for entropies one usually does not need to worry about the process of forming compounds from elements.   This is because *in the limit as T approaches $0°K$, for almost any process which begins with pure reactants and ends with pure* *products, ΔS equals zero.* This generalization from experience is sometimes called the **third law of thermodynamics.** It means, for example, that exactly the same entropy values will be found if dS is integrated for each step of the following two processes: (1) Start with 44 g of crystalline CO_2 at $0°K$ and heat it up to $20°C$. (2) Start with 12 g of carbon and 32 g of oxygen, each pure crystals at $0°K$; heat them and finally react them to give CO_2 at $20°C$. The entropy of formation of almost all pure crystals from their compounds (in contrast to the enthalpy of formation) at $0°K$ is zero. This is certainly convenient, because it saves measuring entropy changes for a large number of chemical reactions in the way that enthalpy changes must be measured. One can just set $S(0°K) = 0$ in Eq. 21-28 for the compound of interest and never worry about making the compound from the elements.

One unfortunate feature of the third law is that in the simple way we have stated it, *it has a number of exceptions.* Thus, while ΔS_f^0 of CO_2 is zero at $0°K$, ΔS_f^0 of CO at $0°K$ is about 1 eu/mole! Other compounds, such as NO, N_2O, CH_3D (where D means deuterium), and even hydrogen have residual entropies at absolute zero. The origin of the third law and of its exceptions lies in the molecular picture of matter, and the reader is referred to F. C. Andrews, *Equilibrium Statistical Mechanics,* John Wiley & Sons, New York, 1963, Section 10, for a brief discussion, or to R. Fowler and E. A. Guggenheim, *Statistical Thermodynamics,* Cambridge University Press, Cambridge, 1939, pp. 191-229, for an extensive discussion.

There is one other statement that is often called the third law of thermo-dynamics, namely the generalization by **Nernst** that *no finite series of processes can lead to a temperature of $0°K$.* This is tantamount to saying that *the absolute zero of temperature is unattainable.* The unattainability of $0°K$ is a physical fact with no exceptions known or expected.* The property $\Delta S(T) \rightarrow 0$ as $T \rightarrow 0°K$ for all processes has exceptions (unless one is very fussy about the meaning of "equilibrium"), and these are well understood from a molecular viewpoint. There are a number of proofs of the identity of the two statements of the third law, but their validity is questionable.

Tables also give Gibbs free energies of formation ΔG_f^0, obtained from

$$\Delta G_f^0 = \Delta H_f^0 - TS^0, \tag{30-50}$$

but in addition, and for the reasons discussed in connection with the enthalpy function, they commonly give values of the **free energy function** (abbreviated

---

* However, temperatures of 10^{-3} $°K$ are regularly achieved in many laboratories and 10^{-5} $°K$ has been reached. So-called *negative temperatures* have also been achieved, but they represent nonequilibrium states which are excessively energetic, that is, very hot (in a sense), and thus are not attained by passing through $0°K$ (see, for example, F. C. Andrews, *op. cit.*, Section 34).

fef), $[G^0(T) - H^0(T^0)]/T$, or else of the negative of this quantity, so that the less precise $H^0(T^0)$ values are separated out and interpolation is easy. The calculation of equilibrium constants from free energy functions is straightforward:

$$- R \ln K = \frac{\Delta G^0}{T} = \sum_i v_i \left[\frac{G^0(T) - H^0(T^0)}{T} \right] + \sum_i v_i \frac{\Delta H^0_{f,i}(T^0)}{T} \quad (30\text{-}51)$$

$$- R \ln K = \Delta \left[\frac{G^0(T) - H^0(T^0)}{T} \right] + \frac{\Delta H^0(T^0)}{T}. \quad (30\text{-}52)$$

The standard enthalpy of reaction term in Eq. 30-52 is the least precise, and gets modified as new ΔH^0 data become available. The free energy function data are often obtained from spectroscopic data with remarkable precision.

When one cannot find the data he needs in tables, one must estimate them from his knowledge of molecular structure. There are a variety of estimation methods involving the contributions of typical groups, valence bond contributions, and a host of other empirical correlations.*

EFFECT OF CHEMISTRY ON THE PHASE RULE

Gibbs' phase rule,

$$f = c - \pi + 2 \quad (30\text{-}53)$$

was derived in Eq. 26-1 for systems without chemical reactions. The definitions of the number of phases π and of the number of degrees of freedom† f are unchanged when chemical reactions are included. In the absence of chemical reactions, each distinct chemical species present is called a component, but if one wants to preserve the form of Eq. 30-53, one must change that definition when reactions occur. This is because for each independent chemical reaction occurring in the system, there is a relationship of the form $\Delta G^0 = -RT \ln K$ which must hold among the intensive variables. Thus in order to obtain the number of components c in a chemically reacting system,

---

* These are discussed extensively, with numerous literature references, in the following books: R. C. Reid and T. K. Sherwood, *The Properties of Gases and Liquids, Their Estimation and Correlation*, McGraw-Hill Book Co., New York, 1958: G. J. Janz, *Thermodynamic Properties of Organic Compounds: Estimation Methods, Principles and Practice*, rev. ed., Academic Press, New York, 1967; W. L. Jolly, *The Synthesis and Characterization of Inorganic Compounds*, Prentice-Hall, Englewood Cliffs, N.J., 1970.

† f is the number of intensive variables used to characterize the system minus the number of relations or restrictions connecting them.

one must subtract from the number of distinct chemical species the number of independent chemical reactions.†

There is another kind of chemical restriction which can reduce the value of c. Suppose in the reaction $A = B + C$ that the only sorce of both B and C is the dissociation of a initial charge of A; thus the concentrations of B and C are equal. There are three chemical species, one equilibrium constant, but only one independently varying component. Whenever such restrictions arise in a system to reduce the number of independently variable mole numbers, the value of c must be reduced accordingly.

EQUILIBRIUM CALCULATIONS—GASES

In a chemically reacting system, for example, $A + B + \cdots = L + M + \cdots$, where all constituents are gases, each activity is the fugacity:

$$K = \frac{f_L f_M \cdots}{f_A f_B \cdots} = \frac{\varphi_L \varphi_M \cdots}{\varphi_A \varphi_B \cdots} \cdot \frac{p_L p_M \cdots}{p_A p_B \cdots}. \qquad (30\text{-}54)$$

We presume that K is available from ΔG^0 and the relationships among the equilibrium partial pressures are our main interest. The activity coefficients are a nuisance, because they have rarely been adequately measured as functions of composition. For moderate pressures (for most organic gases up to 3 atm, in favorable cases as high even as 50 atm), the φ's are close enough to unity to permit using

$$K = K_p \equiv \frac{p_L p_M \cdots}{p_A p_B \cdots} \qquad (p \text{ low}). \qquad (30\text{-}55)$$

For intermediate pressures (up to 100 to 200 atm in favorable situations), the Lewis fugacity rule, Eq. 28-18, has been used with some success. This says that φ_i equals the value of φ_i in *pure* gaseous i at the same temperature and total pressure as the equilibrium mixture. For high pressures, one must find the fugacity coefficients by a more precise means.

Sometimes the partial pressure p_i is a less convenient concentration variable than either the molar concentration $c_i \equiv n_i/V$ or the mole fraction

---

† Determining the number of independent chemical reactions is not always trivial. The three reactions $A + B = C$, $C = 2D$, and $A + B = 2D$ are clearly not independent, only two of them are, and there are only two independent equilibrium constants arising in the system of A, B, C, and D. A thorough discussion of more complicated situations is given by K. Denbigh, *The Principles of Chemical Equilibrium*, Cambridge University Press, London, 1955, section 4.16.

$y_i \equiv p_i/p$. The easiest way to introduce these quantities is, just as needed, to replace the appropriate p_i in Eq. 30-55, thus:

$$p_i \equiv y_i p, \qquad (30\text{-}56)$$

$$p_i \equiv \frac{n_i RT}{V} \equiv c_i RT. \qquad (30\text{-}57)$$

The ideal-gas law was used in obtaining Eq. 30-57. One sometimes encounters use of equilibrium quotients expressed in terms of mole fractions K_y (sometimes written K_x) or concentrations K_c:

$$K_y \equiv \frac{y_L y_M \cdots}{y_A y_B \cdots}, \qquad K_c \equiv \frac{c_L c_M \cdots}{c_A c_B \cdots}, \qquad (30\text{-}58)$$

and

$$K = K_p = K_y p^{\Delta n} = K_c (RT)^{\Delta n}, \quad \text{(gases, p low)} \qquad (30\text{-}59)$$

where Δn is the change in total number of moles of gases in going from left to right in the reaction as written. Since in gas reactions K_p must be independent of total pressure p, K_y must depend on p, as shown by Eq. 30-59. If there are more moles of product gases than reactant gases, Δn is positive, and increasing p decreases K_y, and conversely. Again, as in connection with Eqs. 30-35 and 30-36, the name of Le Chatelier comes to mind.

Suppose one is studying the dissociation of a gaseous compound by heating mass m of it in a volume V and determining p as a function of T. The **degree of dissociation** α is defined as the *fraction* of the molecules of the compound which are dissociated at a given T. Let the stoichiometry of the dissociation be such that one mole of the compound yields v moles of dissociation product. For example, $H_2O = H_2 + \frac{1}{2}O_2$ has $v = \frac{3}{2}$. Every mole of H_2O that was originally put into the vessel is there in the form of $1 - \alpha$ moles of H_2O, α moles of H_2, and $\frac{1}{2}\alpha$ moles of O_2, so the total is $v = \frac{3}{2}$. Similarly, the reaction $N_2O_4 = 2 NO_2$ has $v = 2$; each mole of N_2O_4 in the original charge becomes $1 - \alpha$ moles of N_2O_4 and 2α moles of NO_2 in the equilibrium mixture at T. In general, every mole of the original compound is present at equilibrium as $1 - \alpha$ moles still undissociated plus $v\alpha$ moles of gaseous dissociation product. The total number of moles of original compound put into the vessel is m/M, where M is its molecular weight. The total number of moles of gas at equilibrium is $(1 - \alpha + v\alpha)m/M$, and from the ideal-gas law,

$$pV = \frac{m}{M}(1 - \alpha + v\alpha)RT. \qquad (30\text{-}60)$$

This gives an experimental means of measuring α.

Example 30-4. If α is measured for the reaction $H_2O = H_2 + \frac{1}{2}O_2$ using Eq. 30-60, express α as a function of p, V, T, and m. Also express K_p as a function of α and p.

Solution: First, we solve Eq. 30-60 for α, using $v = \frac{3}{2}$, and get

$$\alpha = \frac{2MpV}{mRT} - 2. \tag{30-61}$$

Next we must find

$$K_p = \frac{p_{H_2} p_{O_2}^{1/2}}{p_{H_2O}} = \frac{y_{H_2} y_{O_2}^{1/2}}{y_{H_2O}} p^{1/2}. \tag{30-62}$$

The mole fractions are obtained immediately. At equilibrium each mole of original H_2O exists as $1 - \alpha$ moles of undissociated H_2O, α moles of H_2, and $\frac{1}{2}\alpha$ moles of O_2. The total number of moles is $(1 - \alpha + \alpha + \frac{1}{2}\alpha) = 1 + \frac{1}{2}\alpha$, so $y_{H_2O} = (1 - \alpha)/(1 + \frac{1}{2}\alpha)$; $y_{H_2} = \alpha/(1 + \frac{1}{2}\alpha)$; and $y_{O_2} = \frac{1}{2}\alpha/(1 + \frac{1}{2}\alpha)$. Thus, Eq. 30-62 becomes

$$K_p = \frac{\alpha^{3/2} p^{1/2}}{2^{1/2}(1 - \alpha)(1 + \frac{1}{2}\alpha)^{1/2}} \tag{30-63}$$

The solution of problems involving gaseous equilibria at low pressures becomes easier as one gains experience and confidence. A wide variety of problems might be encountered, most of which are simply exercises in algebra.

EQUILIBRIUM CALCULATIONS—CONDENSED PHASES

In the chemically reacting system, $A + B + \cdots = L + M + \cdots$, the equilibrium constant is

$$K = \frac{a_L a_M \cdots}{a_A a_B \cdots}. \tag{30-64}$$

If some of the reactants or products are present as either a *pure liquid or pure solid* phase, those substances would be in their standard states if p were 1 atm. Since the change of μ with p is small for condensed phases, at all moderate pressures the activities of all such substances can be set equal to unity.

Example 30-5. Tabulations (from Lewis, Randall, Pitzer, and Brewer, *op. cit.*, pp. 676 and 682) show $-(G^0 - H_{298}^0)/T = 34.47$ cal/deg for $CaCO_3$, 15.26 cal/deg for CaO, and 54.11 cal/deg for CO_2 at 1000°K. The ΔH_{298}^0

values are -288.45, -151.79, and -94.052 kcal, respectively. Calculate the pressure of CO_2 in a mixture of pure $CaCO_3$ and CaO at $1000°K$.

Solution: $\quad CaCO_3 = CaO + CO_2.$ (30-65)

$$\Delta G^0 = -RT \ln p_{CO_2}$$ (30-66)

$$\ln p_{CO_2} = -\frac{\Delta G^0}{RT} = \frac{1}{R} \Delta \left[\frac{-(G^0 - H^0_{298})}{T} \right] - \frac{1}{R} \frac{\Delta H^0_{298}}{T},$$ (30-67)

$$\log p_{CO_2} = \frac{1(\text{deg-mole})}{(2.3)(1.987 \text{ cal})} (15.26 + 54.11 - 34.47$$

$$+ \frac{151.79 + 94.052 - 288.45}{1000}\Bigg) \left(\frac{\text{cal}}{\text{deg-mole}}\right),$$ (30-68)

$$\log p_{CO_2} = -1.69 = -2 + 0.31,$$ (30-69)

$$p_{CO_2} = 0.02 \text{ atm.}$$ (30-70)

Thus if the pressure of CO_2 is maintained above 0.02 atm at $1000°K$, there will be complete conversion of CaO to $CaCO_3$. If the CO_2 pressure is maintained below 0.02 atm, the carbonate is completely converted into the oxide.

In the general equilibrium, $A + B + \cdots = L + M + \cdots$, it is important to realize that the value of K depends upon choice of standard state for each reactant and product. In Eq. 30-64, a_i may be replaced by $\gamma_i x_i$ if the mole fraction scale was chosen for i; it may be replaced by $\gamma_i m_i$, where m means molality, if the molality scale was selected. The different reactants and products might well have different choices made, depending on convenience. For example, in the formation of urea in aqueous solution, $CO_2(g) + 2NH_3(g) = CO(NH_2)_2(aq) + H_2O(l)$, one can choose as standard states unit fugacity for the gases, pure liquid for the water, and the hypothetical ideal solution (in the sense of Henry's law) at unit molality for the urea. The equilibrium constant, then, has the form

$$K = \frac{\gamma_u m_u \gamma_{H_2O} x_{H_2O}}{\varphi_{CO_2} p_{CO_2} \varphi^2_{NH_3} p_{NH_3}}.$$ (30-71)

At low gas pressure and dilute solution, the γ's and φ's ≈ 1 and also $x_{H_2O} \approx 1$. Then $K \approx m_u/p_{CO_2}p_{HN_3}{}^2$. For constituents of condensed phases, it is often most convenient to choose the pressure in the standard state to be p, rather than 1 atm.

Where hydrates of dissolved molecules are formed under such conditions that it is impossible to distinguish between hydrate and nonhydrate, the most useful concentration variable is the *total* molality, that is, the sum of the molality of the hydrate and of the nonhydrate. For example, when ammonia is dissolved in water, the equilibrium $NH_3 + H_2O = NH_4OH$ is established. The total molality is just $m = m_{NH_3} + m_{NH_4OH}$, the stoichiometric concentration of ammonia in the solution (neglecting ionization to form NH_4^+ and OH^-, a phenomenon which *can* be monitored).

The treatment of electrolyte solutions, that is, those in which some of the chemical species are charged ions, parallels in many ways the treatment of nonelectrolytes. Many salts completely dissociate when they dissolve in water, and thus are called **strong electrolytes**; for example, $AgCl(s) = Ag^+(aq) + Cl^-(aq)$. The activity of pure solid AgCl can be taken as unity, so K has the value

$$K = K_{sp} = a_{Ag^+} a_{Cl^-} \equiv \gamma_{\pm}^2 m_{Ag^+} m_{Cl^-}, \tag{30-72}$$

where the equilibrium constant is called the **solubility product constant** K_{sp}. The product of activity coefficients for the two ions is often written $\gamma_{\pm}^2$, because it is hard to measure activity coefficients for each ion separately.

Electrochemical cells can be used to measure ion activities for molalities greater than about 0.05. For more dilute solutions, Gibbs-Duhem integrations are often used, the activity of the solvent being measured by one of the methods previously studied: boiling point elevation, freezing point depression, osmotic pressure, or vapor pressure lowering. Since two ions are formed for every AgCl molecule that dissolves, electrolytes have correspondingly greater effects on solvent activities than would otherwise be expected because their mole fraction increases at a multiple of their simple stoichiometric mole fraction. At very low concentrations where molalities and mole fractions are proportional, the activity coefficient represents departures from ideality. At moderate concentrations even ideal solutions would show departures from $\gamma = 1$ because of the choice of the molality scale. However, ideality (in the sense of Henry's law) is restricted in ionic solutions to extraordinarily dilute solutions, say, less than $10^{-4} m$, while nonelectrolytes are commonly ideal up to perhaps 0.05 molal. The reason for the difference lies in the very large distances over which the coulombic forces of the charged ions are effective. A solution of "infinite dilution" must be much more dilute when the particles are charged than when they are not. The higher the charge on ions in the solution, the greater the departures from $\gamma_{\pm} = 1$. These nonideality effects are most important, as the following $\gamma_{\pm}$ values for 1 molal solutions at 25°C show: for HCl, $\gamma_{\pm} = 0.8$; for H_2SO_4, $\gamma_{\pm} = 0.15$; for $MgSO_4$, $\gamma_{\pm} = 0.07$; and for $Al_2(SO_4)_3$, $\gamma_{\pm} = 0.018$. For high concentrations, equally dramatic departures from ideality can arise, for example, 16 molal HCl has $\gamma_{\pm} = 41.5$. These $\gamma_{\pm}$ values are especially significant in view of the fact that $\gamma_{\pm}$ is raised to the power of the number of ions formed from a molecule of undissociated electrolyte. For $Al_2(SO_4)_3$, for examle, $\gamma_{\pm}^5 = 1.9 \times 10^{-9}$!

The ionization of NH_4OH is typical of the behavior of weak electrolytes: $NH_4OH = NH_4^+ + OH^-$. Here,

$$K = \frac{a_{NH_4^+} \, a_{OH^-}}{a_{NH_4OH}} = \frac{\gamma_\pm^2 m_{NH_4^+} \, m_{OH^-}}{\gamma_{NH_4OH} \, m_{NH_4OH}} \tag{30-73}$$

where by the molality of NH_4OH is meant the sum of the NH_4OH and the NH_3 molalities in the solution, as mentioned above.

When the Van't Hoff equation is written for the process of dissolving a salt (like AgCl above), the ΔH^0 that enters is the heat of solution of the salt at infinite dilution. For the ionization of a weak electrolyte, ΔH^0 is the heat of ionization at infinite dilution. The approximation that ΔH^0 is temperature independent is particularly poor for ionization processes, for which Δc_p is often around 40 cal/deg-mole.

As the reader is no doubt aware, a variety of complicated problems arise when several independent ionic equilibria occur in a single solution. Since solvents themselves are always slightly ionized (for example, $K_w = a_H \, a_{OH^-} \approx 10^{-14}$ at 25°C), several equilibria are present in almost all solutions of interest. A common approach to all these problems facilitates their solution and removes the anxiety of trying to remember the right special trick to solve the particular one:

1. Write down all species whose activity plays a role in the chemical processes considered. As an example, in the preparation of a buffer from sodium acetate (abbreviated NaAc) and acetic acid these are

$$H^+, OH^-, Na^+, Ac^-, HAc \tag{30-74}$$

(water is neglected since the solution is presumed sufficiently dilute for its activity to be nearly constant).

2. Write down the equilibrium constant expression for each independent chemical reaction in the system. In this example these are the ionization of H_2O and of HAc:

$$K_w = [H^+][OH^-], \qquad K_a = \frac{[H^+][Ac^-]}{[HAc]}. \tag{30-75}$$

The conventional bracket notation is used for molality, and the activity coefficients, if required, have been incorporated into the K's.

3. Write down the electroneutrality condition, assuring that the concentrations of positive and negative charges be equal:

$$[H^+] + [Na^+] - [OH^-] - [Ac^-] = 0. \tag{30-76}$$

The left-hand side of Eq. 30-76 is the sum of all the species concentrations from Eq. 30-74 which bear charges, each one multiplied by its charge.

4. We now have fewer equations, Eqs. 30-75 and 30-76, among the variables of Eq. 30-74 than we have variables. There will always be as many independent mass balances among the species as are needed to make the number of equations equal the number of species. In our example, suppose we made up the solution by adding C_s moles of the salt NaAc to each thousand grams of final solution; thus C_s is the **stoichiometric concentration** (in contrast to the equilibrium concentration) of NaAc. Similarly, let C_a be the stoichiometric concentration of HAc. Since we are not treating $[H_2O]$ as a variable, we cannot perform mass balances on water, hydrogen, or hydroxide. This leaves only sodium and acetate from the list, Eq. 30-74:

$$[Na^+] = C_s, \qquad [HAc] + [Ac^-] = C_a + C_s. \qquad (30\text{-}77)$$

The left-hand sides of Eqs. 30-77 are the equilibrium concentrations of Na^+ and Ac; the right-hand sides are the stoichiometric concentrations. Since matter is conserved, these must be equal.

We now have as many equations as chemical species. These equations couple the set of variables which consists of the equilibrium concentrations of all the species plus the stoichiometric concentrations that were introduced in the mass balances. In a particular application, certain of these variables will be fixed and the rest can be solved for. Judicious approximations can usually be made to simplify things (for example, either $[OH^-] \ll [H^+]$ or $[OH^-] \gg [H^+]$), but at least one starts with exact equations and thus can study the response of particular quantities to other variables *over a range* of variation of the latter.

SUGGESTED READING

Butler, J. D., *Ionic Equilibrium*, Addison-Wesley Publishing Company, Reading, Mass., 1964.

PROBLEMS

30-1. Prove the assertion that molality is proportional to mole fraction for very dilute solutions. At what molality in a water solution does the molality of the solute (2) depart from proportionality to x_2 by 1% of its value?

30-2. Use the numerical values of R, T and $\mathcal{F}$, converting $\ln K$ to $\log K$ in Eq. 30-25 to write the Nernst equation at 25°C.

30-3. The following absolute molar entropies are given at 25°C: for $H_2(g)$, 31.211 eu; for $Cl_2(g)$ 53.286 eu; for $HCl(g)$, 44.617 eu. What is the molar entropy of formation of $HCl(g)$ from $H_2(g)$ and $Cl_2(g)$ at 25°C? What is $\Delta S°$ for the reaction $2HCl(g) = H_2(g) + Cl_2(g)$?

30-4. For the reaction $N_2 + 3H_2 = 2NH_3$ at 400°C, $K_p = 1.64 \times 10^{-4}$ atm^{-2}. Calculate K_c and also calculate K_y at 10 atm.

30-5. Calculate the enthalpy change for the reaction $N_2(g) + O_2(g) = 2NO(g)$ from knowledge that at 2000°K the equilibrium constant is 4.08×10^{-4} and that at 2500°K it is 36.0×10^{-4}.

30-6. At 26°C the Gibbs free energy of formation of normal butane is -3.754 kcal/mole. For isobutane it is -4.296 kcal/mole. In an equilibrium mixture of these gases at 25°C, what fraction of the molecules will be isobutane?

30-7. PCl_5 dissociates to form PCl_3 and Cl_2; all are gases. At 250°C, 1 l of partially dissociated PCl_5 at 1 atm weighs 2.690 g. Find α and K_p.

30-8. Use the result of Prob. 30-7 to find α at 250°C if 0.1 mole of PCl_5 is placed in a 3-l vessel containing Cl_2 gas already at a pressure of 0.5 atm.

30-9. How many moles of PCl_5 must be put into a liter vessel at 250°C to get a Cl_2 concentration of 0.1 mole/l? Use the results of Probs. 30-7 and 30-8 if necessary.

30-10. Under what total pressure must an equimolar mixture of Cl_2 and PCl_3 be placed to get a partial pressure of PCl_5 of 1 atm at 250°? Use the results of Probs. 7-9, if necessary.

30-11. Under what pressure must an equimolar mixture of Cl_2 and PCl_3 be at 250°C to get 80% conversion of PCl_3 into PCl_5? Use the results of Probs. 7-10, if necessary.

30-12. At a given temperature and pressure a diatomic gas is 0.1% dissociated into atoms. What will be the degree of dissociation if the pressure is lowered by a factor of 100?

30-13. A 0.25 molal solution of $ZnSO_4$ in water freezes at -0.62°C. The molal freezing point depression of water is -1.86°C. What is the apparent equilibrium constant in molalities for the dissociation of $ZnZO_4$? How important would you guess nonideality to be in this experiment?

30-14. The heat of combustion of ethylene (C_2H_4) is 336.6 kcal/mole; that of cyclohexane (C_6H_{12}) is 937.8 kcal/mole. What is the heat of polymerization of ethylene to cyclohexane? Is the heat evolved or absorbed?

30-15. At 25°C, the heat of dissociation of nitrogen gas is 225,900 cal/mole, and the heat of formation of ammonia gas is $-10,950$ cal/mole. What is the change of internal energy on dissociation of 1 mole of ammonia gas into gaseous atoms? The enthalpy of formation of gaseous H atoms at 25°C is 52.089 kcal/mole.

30-16. The equilibrium constant for the reaction $H_2 = 2H$ is 7.0×10^{-18} atm at 1000°K and 3.1×10^{-6} atm at 2000°K. Estimate the heat of dissociation of H_2.

30-17. For strong electrolytes such as HCl in solution, the chemical potential of the HCl is often equated to the sum of the chemical potentials of the individual ions. What is the equivalent expression to Henry's law for dilute solutions of HCl gas in water (use molalities as solution concentration variable)?

30-18. Estimate the maximum flame temperature for the combustion of H_2 in Cl_2. The initial temperature is 25°C. The enthalpy of formation of $HCl(g)$ at 25°C is -22.1 kcal/mole.

30-19. It has been said that a two-minute egg in New York City is a four-minute egg in Denver, where the atmospheric pressure is about 0.816 atm. The heat of vaporization of water is 595 cal/g; what is ΔH^0 for the protein denaturation involved in the boiling of an egg?

30-20. A reaction has $\mathscr{E}^0 = +1.330$ volts at 300°K and $\mathscr{E}^0 = +1.325$ volts at 310°K. What is ΔS for the reaction, if two electrons are involved in the redox process of the reaction as written?

30-21. Calculate ΔH at 170°C for the reaction $CdSO_4(s) + H_2O(g) = CdSO_4 \cdot H_2O(s)$ using the following information at 20°C:

$$CdSO_4(s) + 400H_2O(l) = CdSO_4 \text{ in } 400 \text{ } H_2O \qquad \Delta H = -10,900 \text{ cal}$$
$$CdSO_4 \cdot H_2O(s) + 399H_2O(l) = CdSO_4 \text{ in } 400H_2O \qquad \Delta H = -6,100 \text{ cal}$$
$$H_2O(g) = H_2O(l) \qquad \Delta H = -9,700 \text{ cal}$$

Assume that $H_2O(g)$ has $c_p = 7.2$ cal/deg-mole, $CdSO_4(s)$ has $c_p = 41.6$ cal/deg-mole, and $CdSO_4 \cdot H_2O(s)$ has $c_p = 46.6$ cal/deg-mole.

30-22. As long as electric power is generated from heat engines, thermodynamics dictates the amount of termal pollution which must accompany the power production (for example, see Prob. 11-13). Suppose the oxidation of the fuel could be accomplished in a so-called *fuel cell*, a device in which the ΔG of reaction is directly generated as electric power at constant temperature. Of course, Q for the reversible process is $T \Delta S$, so a reaction with ΔS positive is preferred to one where it is negative. A fuel cell has been made which harnesses the reaction $2H_2(g) + O_2(g) = 2H_2O(g)$, for which $\Delta G^0 = -109.28$ kcal and $\Delta H^0 = -115.60$ kcal at 25°C. What is Q for the reaction as written? How much work would be generated? How does this compare with Q and W for burning the hydrogen in a conventional heat engine with boiler temperature 500°C and condenser temperature 25°C?

30-23. At 25°C, the half-reaction $Mg = Mg^{++} + 2e^-$ has $\mathscr{E}^0 = 2.37$ volts, the reaction $Mg + 2OH^- = Mg(OH)_2 + 2e^-$ has $\mathscr{E}^0 = 2.69$ volts. Calculate K_{sp}, the solubility product constant for $Mg(OH)_2$.

30-24. Sulfuric acid ionizes according to the following scheme: $H_2SO_4 \rightarrow H^+ + HSO_4^-$ goes to completion; $HSO_4^- = H^+ + SO_4^{--}$ has $K = 1.2 \times 10^{-2}$; and H_2O of course has $K_w = 10^{-14}$. Solve for the stoichiometric concentration of H_2SO_4, C (the number of moles of H_2SO_4 added to water to form a liter of solution), as a function of (H^+), K, and K_w. *Note*: (H^+) as function of C is then obtained immediately from the graph of C as function of (H^+).

30-25. The salt of a weak acid hydrolyzes as follows: $A^+ + H_2O + B^- = AOH + H^+ + B^-$. The reaction $AOH = A^+ + OH^-$ has ionization constant $K_B = (A^+)(OH^-)/(AOH)$. The stoichiometric concentration of salt is C_s. Express C_s as function of (H^+), K_B, and K_w. What is the form of this at moderate C_s?

30-26. The stoichiometric concentration of KI is C_1, of $Fe(NO_3)_2$ is C_2, of $Fe(NO_3)_3$ is C_3, and an excess of I_2 crystals is present. Find the equilibrium concentrations of the various ions. The reaction $Fe^{++} + \frac{1}{2}I_2 = Fe^{3+} + I^-$ has equilibrium constant $K = (Fe^{3+})(I^-)/(Fe^{++})$.

APPENDIX A: CALCULUS OF SEVERAL VARIABLES

FUNCTIONS OF SEVERAL VARIABLES

This appendix discusses some features of the calculus of several variables that recur again and again in thermodynamics. Most of science is concerned with functions of several variables. When one says that something, A, is a *function* of something else, B, one means that there exists some kind of recipe for finding A once B is given. Thus when we say the pressure is a function of the state of the system, we mean there is a recipe for finding the pressure once the state of the system is specified. This recipe is given by nature. Consider the case of a large, fixed amount of material in an equilibrium state which is specified by the internal energy U and the volume V. Thus all equilibrium functions of state are functions of the variables U and V for this material. This means, for example, that nature affords a means for obtaining the pressure for this system as a function of U and V. The recipe is to take the specified material, give it the appropriate U and V, and let it reach equilibrium. The pressure, defined as the force per unit area in the system, is then measured by an appropriate technique. It thus is *nature* which provides the recipe for the functional relation.

Of course, there are ways to avoid having to repeat an experiment every time a value of pressure is wanted. Tables of pressures for various choices of U and V may be constructed from a set of experiments. Or graphs and curves showing how pressure depends on the state variables may be drawn. It is convenient to approximate these tables and graphs by mathematical relations showing how to get the pressure from U and V by arithmetical manipulation. These mathematical functional relations are always approximations, chosen for their convenience, to the actual recipes given by nature.

Thus when we say the pressure p is a function of U and V,

$$p = p(U, V), \tag{A-1}$$

we usually are thinking of a mathematical approximation to nature's functional dependence. The left-hand side of Eq. A-1 is the symbol for the pressure.   The right-hand side is some mathematical expression involving only the symbols for internal energy and volume (plus, of course, various constants where appropriate).   One says that Eq. A-1 gives p as an **explicit function** of U and V; the recipe for finding p from U and V is completely specified.   Sometimes Eq. A-1 is written symmetrically, with p not singled out, by subtracting the right-hand side from both sides:

$$p - p(U, V) = 0, \qquad (A\text{-}2)$$

or, defining the left-hand side of this by $f(p, U, V)$, a new function of the three variables p, U, and V:

$$f(p, U, V) = 0. \qquad (A\text{-}3)$$

For the kinds of functions encountered in thermodynamics, it is almost always possible to view equations like Eq. A-3 as *implying* functional dependence of *any one* of the variables, p, U, V on the rest.   For example, Eq. A-3 could be solved for U to yield the explicit function

$$U = U(p, V). \qquad (A\text{-}4)$$

All four of the foregoing equations express the same relationship among the variables U, p, and V; Eq. A-1 gives p as an explicit function of U and V; Eq. A-4 gives U as an explicit function of p and V; Eq. A-3 is a relation implicitly defining all three relations: $p = p(U, V)$, $U = U(p, V)$, and $V = V(p, U)$.

Not only is the pressure a function of the state of the system but also all the other system properties, such as the temperature, are also functions of the state variables:

$$T = T(U, V). \qquad (A\text{-}5)$$

If Eq. A-4 is used to eliminate the variable U from Eq. A-5, the temperature becomes an explicit function of p and V, rather than of U and V.

$$T = T(p, V). \qquad (A\text{-}6)$$

While scientists are used to seeing equations like A-5 and A-6 side by side, most mathematicians would be appalled by the sight.   They would use the T on the right side of Eq. A-5 to represent a particular mathematical recipe for going from numbers representing U and V in some appropriate units to a number representing the temperature in some determined units.   Obviously, if T is treated as an explicit function of p and V, as it is in Eq. A-6, the mathematical recipe must be different from that of Eq. A-5, and mathematicians would use a different symbol.   On the other hand, scientists are likely to *use the symbol to represent the property*, regardless of what variables it is being viewed as a function of.   This helps the scientist to keep a physical feeling for the terms in his equations. The T on the right side of Eq. A-5 means a particular relationship among U and V that approximates the temperature; the T on the right side of Eq. A-6 means a completely different relationship among p and V that for the same system approximates the temperature. Of course, if scientists tried to use different symbols for each function, they would soon run out of alphabet, including Greek, Russian, and Hebrew. Mathematicians usually deal with relatively few " quantities " and with only a single change of variables, not with all combinations of p, S, T, V, U, H, A, G, n_1, μ_1, n_2, μ_2, ...

The expressions of thermodynamics often involve more than two independent variables. All the statements made in this appendix apply to functions of any number of variables. Functions of two are easier to visualize pictorially; that is why they have been chosen. The functions encountered in thermodynamics are almost always continuous, that is, the smaller the changes in independent variables U and V, the smaller the changes in the dependent quantity, in this case, p. That means that if U and V in appropriate units are plotted in Cartesian coordinates, the points $p(U, V)$ form a continuous surface in p-U-V space, as shown in Fig. A-1. The surface may

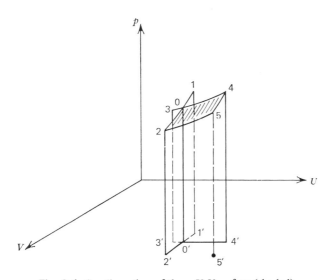

Fig. A-1. Small portion of the p-U-V surface (shaded).

be viewed as being the locus of all points in p-U-V space for which all of the equivalent equations, Eqs. A-1 through A-4, are satisfied. The equation coupling the variables p, U, and V restricts the allowed points in p-U-V space for the given equilibrium system to those on the surface.

PARTIAL DERIVATIVES

Through the point 0 in Fig. A-1, planes perpendicular to the U and the V axes have been drawn. This makes it easy to see the two principal slopes of the $p(U, V)$ surface. The plane 3044′0′3′ is a surface of constant V. The line 304 is the locus of points satisfying Eq. A-1 through A-4 and having this particular value of V. The slope of the line 304 is the rate of change of p with changing U, holding V fixed at this value. This slope is written $(\partial p/\partial U)_V$, the V subscript showing what variable is held constant. Similarly, 1022′0′1′ is a surface of constant U, and the slope $(\partial p/\partial V)_U$ of the line 102 is the rate of change of p with increasing V, holding U at this fixed value.

To find the change dp in a dependent variable p for infinitesimal changes dV and dU in the independent variables V and U, the following expression is used:

$$dp = \left(\frac{\partial p}{\partial U}\right)_V dU + \left(\frac{\partial p}{\partial V}\right)_U dV. \tag{A-7}$$

This has the following geometric interpretation: For any except the most pathological surfaces where continuity requirements are violated, a sufficiently small region of the surface (like 0254 in Fig. A-1) may be approximated by a plane. Clearly, on this plane, Eq. A-7 is rigorously true. In Fig. A-1, for example, the analog to Eq. A-7 is

$$(p_5 - p_0) = \left(\frac{\partial p}{\partial U}\right)_V (U_4 - U_0) + \left(\frac{\partial p}{\partial V}\right)_U (V_2 - V_0), \tag{A-8}$$

which computes the change in p in two steps. The first is that due to changing U and holding V constant. The second is that due to changing V, holding U constant. The small surface differs little from a plane and its edges differ little from straight lines, so in the limit of infinitesimal changes, Eq. A-8 becomes Eq. A-7 and is rigorously true.

A mathematical interpretation of Eq. A-7 is that of a Taylor series expansion of, say, p_B about p_A:

$$p_B = p_A + \left(\frac{\partial p}{\partial U}\right)_V (U_B - U_A) + \left(\frac{\partial p}{\partial V}\right)_U (V_B - V_A) + \frac{1}{2}\left(\frac{\partial^2 p}{\partial U^2}\right)_V (U_B - U_A)^2$$

$$+ \frac{1}{2}\left(\frac{\partial}{\partial V}\right)_U \left(\frac{\partial p}{\partial U}\right)_V (V_B - V_A)(U_B - U_A)$$

$$+ \frac{1}{2}\left(\frac{\partial}{\partial U}\right)_V \left(\frac{\partial p}{\partial V}\right)_U (U_B - U_A)(V_B - V_A) + \frac{1}{2}\left(\frac{\partial^2 p}{\partial V^2}\right)_U (V_B - V_A)^2 + \cdots, \tag{A-9}$$

where the intervals $U_B - U_A$ and $V_B - V_A$ have been shrunk so small that second powers of them may be neglected compared to first powers. In the limit of truly infinitesimal intervals, Eq. A-7 follows rigorously from Eq. A-9.

One point about partial derivatives is the following: Suppose one writes the total differential of, say, $U = U(S, V)$:

$$dU = \left(\frac{\partial U}{\partial S}\right)_V dS + \left(\frac{\partial U}{\partial V}\right)_S dV, \qquad \text{(A-10)}$$

an equation which couples arbitrary infinitesimal changes in U, S, and V. The quantities $(\partial U/\partial S)_V$ and $(\partial U/\partial V)_S$ are functions of the independent variables S and V, obtainable from $U = U(S, V)$ by differentiation. Now suppose one learns from some other source that

$$dU = T\,dS - p\,dV \qquad \text{(A-11)}$$

for arbitrary infinitesimal changes in U, S, and V. Here T and p are functions of S and V. The mere comparison of Eqs. A-10 and A-11 would not imply that

$$T = \left(\frac{\partial U}{\partial S}\right)_V \qquad \text{and} \qquad -p = \left(\frac{\partial U}{\partial V}\right)_S, \qquad \text{(A-12)}$$

were it not for the fact that both Eqs. A-10 and A-11 must hold for *arbitrary* changes in U, S, and V. Just because $x = ay + bz$ and $x = cy + dz$, it does not follow that $a = c$ and $b = d$; it follows only that $(a - c)y + (b - d)z = 0$. However, the fact that they hold for arbitrary changes means they must be equal term by term. As proof, for example, we may let U and S change and keep V constant. Under those conditions, Eqs. A-10 and A-11 yield the two equations:

$$dU = \left(\frac{\partial U}{dS}\right)_V dS, \qquad dU = T\,dS \qquad (V\ \text{constant}). \qquad \text{(A-13)}$$

This proves that T must be the same function of V and S that $(\partial U/\partial S)_V$ is. Similarly, Eqs. A-10 and A-11 become

$$dU = \left(\frac{\partial U}{\partial V}\right)_S dV, \qquad dU = -p\,dV \qquad (S\ \text{constant}). \qquad \text{(A-14)}$$

if S is kept constant, and therefore both of Eqs. A-12 must follow from Eqs. A-10 and A-11.

INTEGRATING TOTAL DERIVATIVES

Clearly, one can find the difference in any function of state like p between two states by summing the infinitesimal change in the function dp over each step of *any* process which connects the two states:

$$\Delta p = p_B - p_A = \int_A^B dp \quad \text{(integral over any path).} \quad \text{(A-15)}$$

There are an infinite number of paths, of course, leading from state A to state B; one chooses a path along which the value of dp is known from Eq. A-7. For example, if $(\partial p/\partial V)_U$ is known as a function of V for $U = U_B$ and if $(\partial p/\partial U)_V$ is known as a function of U for $V = V_A$, then Δp may be found from

$$\Delta p = \int_{U_A, V_A}^{U_B, V_A} \left(\frac{\partial p}{\partial U}\right)_V dU + \int_{U_B, V_A}^{U_B, V_B} \left(\frac{\partial p}{\partial V}\right)_U dV. \quad \text{(A-16)}$$

In both terms of Eq. A-16, dp is given by Eq. A-7; but along the first step, dV is zero; and along the second step, dU is zero.

It is important to realize that just as p is a function of both U and V, so also in general are the functions $(\partial p/\partial U)_V$ and $(\partial p/\partial V)_U$. Even though, say, $(\partial p/\partial U)_V$ is found from p by differentiation holding V constant, $(\partial p/\partial U)_V$ still depends on the value of the constant V.

Another path between states A and B along which the integral of dp must give the same value as Eq. A-16 is U_A, $V_A \to U_A$, $V_B \to U_B$, V_B, which yields

$$\Delta p = \int_{U_A, V_B}^{U_B, V_B} \left(\frac{\partial p}{\partial U}\right)_V dU + \int_{U_A, V_A}^{U_A, V_B} \left(\frac{\partial p}{\partial V}\right)_U dV. \quad \text{(A-17)}$$

GENERAL PROPERTIES OF PARTIAL DERIVATIVES

We now find some general properties of partial derivatives. We rewrite Eq. A-7 as

$$\left(\frac{\partial p}{\partial V}\right)_U dV = dp - \left(\frac{\partial p}{\partial U}\right)_V dU, \quad \text{(A-18)}$$

$$dV = \frac{1}{(\partial p/\partial V)_U} dp - \frac{(\partial p/\partial U)_V}{(\partial p/\partial V)_U} dU. \quad \text{(A-19)}$$

In Eq. A-19 we simply divided through by $(\partial p/\partial V)_U$.* Since the result relates arbitrary changes in p, U, and V, it must be identical to the expression for the total differential of V:

$$dV = \left(\frac{\partial V}{\partial p}\right)_U dp + \left(\frac{\partial V}{\partial U}\right)_p dU. \quad \text{(A-20)}$$

---

* This paragraph is meaningful only if the slopes are not zero.

Since both Eqs. A-19 and A-20 must be valid for arbitrary infinitesimal changes in V, p, and U, we may use the conclusions of Eqs. A-10 to A-14 to equate the coefficients of like derivatives. The coefficients of dp yield

$$\left(\frac{\partial V}{\partial p}\right)_U = \frac{1}{(\partial p/\partial V)_U} \qquad \text{(general property of derivatives)}, \qquad \text{(A-21)}$$

which is intuitively obvious from the meaning of the derivative as a slope. The coefficients of dU yield

$$\left(\frac{\partial V}{\partial U}\right)_p = -\frac{(\partial p/\partial U)_V}{(\partial p/\partial V)_U} = -\frac{(\partial V/\partial p)_U}{(\partial U/\partial p)_V} \qquad \text{(general property of derivatives)}.$$

$$\text{(A-22)}$$

In the last step of Eq. A-22, Eq. A-21 was used on both numerator and denominator. The structures of these equations are simple enough that they can quite easily be remembered; they are very useful. Mistakes in remembering can be caught by checking the dimensions, which must of course be the same on both sides of an equation. The dimensions of p, U, and V are all different, and they must properly cancel if the resulting equation is correct. One sometimes sees Eq. A-22 summarized in the form

$$\left(\frac{\partial V}{\partial U}\right)_p \left(\frac{\partial U}{\partial p}\right)_V \left(\frac{\partial p}{\partial V}\right)_U = -1 \qquad \text{(general property of derivatives).} \quad \text{(A-23)}$$

EQUALITY OF MIXED SECOND DERIVATIVES

One more feature of the derivatives for a surface such as shown in Fig. A-1 is the equality of the mixed second derivatives,

$$\left(\frac{\partial}{\partial V}\right)_U \left(\frac{\partial p}{\partial U}\right)_V = \left(\frac{\partial}{\partial U}\right)_V \left(\frac{\partial p}{\partial V}\right)_U \qquad \text{(general property of derivatives).} \quad \text{(A-24)}$$

In this case, the surface 0254 in Fig. A-1 must not be viewed as planar, otherwise these second derivatives would vanish. The physical meaning of Eq. A-24 is perhaps clarified by thinking through the following argument while considering Fig. A-1: Let us consider the change in value of p between point 5 and point 0. The change in U between these points is $\Delta U = U_4 - U_0 = U_5 - U_2$. The change in V between these points is $\Delta V = V_2 - V_0 = V_5 - V_4$. The change in p over the distance ΔU is $(\partial p/\partial U)_V \Delta U$. At $V = V_0$, this change has the value $p_4 - p_0$; at $V = V_2$, it has the value $p_5 - p_2$. Since the change in any quantity f in the distance ΔV is $(\partial f/\partial V)_U \Delta V$, the change in $(\partial p/\partial U)_V \Delta U$ in $V_2 - V_0$ is

$$\left(\frac{\partial}{\partial V}\right)_U \left[\left(\frac{\partial p}{\partial U}\right)_V \Delta U\right] \Delta V = (p_5 - p_2) - (p_4 - p_0). \qquad \text{(A-25)}$$

Similarly, $(\partial p/\partial V)_U \Delta V$ has the value $p_2 - p_0$ at U_0 and the value $p_5 - p_4$ at U_4. Thus, the change in $(\partial p/\partial V)_U \Delta V$ in the distance $U_4 - U_0$ is

$$\left(\frac{\partial}{\partial U}\right)_V \left[\left(\frac{\partial p}{\partial V}\right)_U \Delta V\right] \Delta U = (p_5 - p_4) - (p_2 - p_0). \qquad \text{(A-26)}$$

The intervals ΔU and ΔV are constants and the right-hand sides of Eqs. A-25 and A-26 are both equal to $p_5 - p_4 - p_2 + p_0$. Thus Eq. A-24 emerges upon cancellation of ΔV and ΔU between the left hand sides of Eqs. A-25 and A-26. The equality of the mixed partial derivatives is thus derived as a consequence of the fact that $p(U, V)$ is an actual surface. Indeed, Eq. A-24 is **Cauchy's criterion** for the integrability of the quantity p, that is, for the existence of a surface of the form $f(p, U, V) = 0$.*

Warning: The cross partial derivatives are equal, Eq. A-24, only if the independent variables are the same in both differentiations. That is,

$$\left(\frac{\partial}{\partial V}\right)_S \left(\frac{\partial p}{\partial U}\right)_V \qquad \text{and} \qquad \left(\frac{\partial}{\partial U}\right)_V \left(\frac{\partial p}{\partial V}\right)_T \qquad \text{(A-27)}$$

have no more likelihood of being equal than any two randomly chosen functions. Interchanging order of such differentiation has led many people to grief.

CHANGE OF INDEPENDENT VARIABLE

The next problem considered here is what happens on changing from one independent variable to another. For example, suppose one knows p as a function of U and V, and one also knows U as a function of T and V. The problem is to change Eq. A-7, which gives dp in terms of explicit changes in U and V to a similar equation giving dp in terms of explicit changes in T and V.

A direct approach will solve this problem. The trouble with Eq. A-7,

$$dp = \left(\frac{\partial p}{\partial U}\right)_V dU + \left(\frac{\partial p}{\partial V}\right)_U dV, \qquad \text{(A-28)}$$

is that it has dU and dV rather than dT and dV. However, since $U = U(T, V)$ dU can be expressed in terms of dT and dV:

---

* For a discussion, see any advanced calculus book, for example. R. Courant, *Differential and Integral Calculus*, Blackie & Son Ltd., London, 1936, vol. II, pp. 55–59.

$$dU = \left(\frac{\partial U}{\partial T}\right)_V dT + \left(\frac{\partial U}{\partial V}\right)_T dV, \tag{A-29}$$

and this expression may be used in Eq. A-28 to obtain

$$dp = \left(\frac{\partial p}{\partial U}\right)_V \left(\frac{\partial U}{\partial T}\right)_V dT + \left[\left(\frac{\partial p}{\partial V}\right)_U + \left(\frac{\partial p}{\partial U}\right)_V \left(\frac{\partial U}{\partial V}\right)_T\right] dV, \tag{A-30}$$

where the coefficients of dT and dV have been isolated. We may compare Eq. A-30 with the expression for the total derivative of $p = p(T, V)$:

$$dp = \left(\frac{\partial p}{\partial T}\right)_V dT + \left(\frac{\partial p}{\partial V}\right)_T dV, \tag{A-31}$$

and, in accordance with Eqs. A-10 to A-14 equate the coefficients of like derivatives. The coefficients of dT yield

$$\left(\frac{\partial p}{\partial T}\right)_V = \left(\frac{\partial p}{\partial U}\right)_V \left(\frac{\partial U}{\partial T}\right)_V \qquad \text{(general property of derivatives),} \tag{A-32}$$

which is merely a chain rule, only for partial derivatives rather than total. Note, in order to use the chain rule, the same quantity must be held constant in each derivative. Then, of course, for all practical purposes one can forget that one is dealing with functions of several variables. Equating the coefficients of dV in Eqs. A-30 and A-31 yields

$$\left(\frac{\partial p}{\partial V}\right)_T = \left(\frac{\partial p}{\partial V}\right)_U + \left(\frac{\partial p}{\partial U}\right)_V \left(\frac{\partial U}{\partial V}\right)_T \qquad \text{(general property of derivatives).}$$

$$\tag{A-33}$$

This last equation permits changing the quantity kept constant in a differentiation.

This may be visualized geometrically in Fig. A-2, which resembles Fig. A-1,

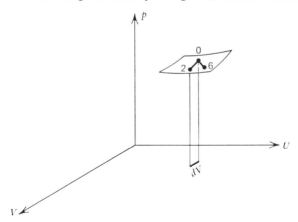

Fig. A-2. Representation of Eq. A-33, line 06 an isotherm (that is, T is constant).

with the addition of the isotherm (line of constant T), line 06, through the point 0. We multiply Eq. A-33 through by dV,

$$\left(\frac{\partial p}{\partial V}\right)_T dV = \left(\frac{\partial p}{\partial V}\right)_U dV + \left(\frac{\partial p}{\partial U}\right)_V \left(\frac{\partial U}{\partial V}\right)_T dV, \tag{A-34}$$

and note that physically the quantity on the left-hand side is

$$\left(\frac{\partial p}{\partial V}\right)_T dV = p_6 - p_0, \tag{A-35}$$

since the change of state is along the isotherm. The quantity represented by the first term on the right-hand side of Eq. A-34 is

$$\left(\frac{\partial p}{\partial V}\right)_U dV = p_2 - p_0, \tag{A-36}$$

since the change is made at constant U. The right-hand part of the last term in Eq. A-34 is

$$\left(\frac{\partial U}{\partial V}\right)_T dV = U_6 - U_0 = U_6 - U_2, \tag{A-37}$$

since $U_2 = U_0$. Therefore the quantity represented by the entire last term of Eq. A-34 is

$$\left(\frac{\partial p}{\partial U}\right)_V \left(\frac{\partial U}{\partial V}\right)_T dV = \left(\frac{\partial p}{\partial U}\right)_V (U_6 - U_2) = p_6 - p_2. \tag{A-38}$$

Thus the sum of Eqs. A-36 and A-38 equals Eq. A-35, and the geometric picture of Eq. A-33 is perhaps clear.

The last property of derivatives that we treat is, first, to derive a general expression which looks terribly complicated, but which is easily interpreted and which often is helpful. We start with the chain rule, Eq. A-32:

$$\left(\frac{\partial U}{\partial x}\right)_y = \left(\frac{\partial S}{\partial x}\right)_y \left(\frac{\partial U}{\partial S}\right)_y. \tag{A-39}$$

We then use Eq. A-33 to change the last derivative:

$$\left(\frac{\partial U}{\partial x}\right)_y = \left(\frac{\partial S}{\partial x}\right)_y \left[\left(\frac{\partial U}{\partial S}\right)_V + \left(\frac{\partial U}{\partial V}\right)_S \left(\frac{\partial V}{\partial S}\right)_y\right], \tag{A-40}$$

$$\left(\frac{\partial U}{\partial x}\right)_y = \left(\frac{\partial U}{\partial S}\right)_V \left(\frac{\partial S}{\partial x}\right)_y + \left(\frac{\partial U}{\partial V}\right)_S \left(\frac{\partial V}{\partial x}\right)_y \qquad \text{(general property of derivatives),} \tag{A-41}$$

where we again used the chain rule in going from Eq. A-40 to A-41.

This result bears some study.   If we compare the expression for the total derivative of U,

$$dU = \left(\frac{\partial U}{\partial S}\right)_V dS + \left(\frac{\partial U}{\partial V}\right)_S dV \tag{A-42}$$

with Eq. A-41, we can interpret A-41 as follows: Eq. A-42 relates arbitrary infinitesimal changes in U, S, and V.   In particular, it relates those changes when x is changing too, whatever x may be, and when y is held constant, whatever y might be.   Thus if Eq. A-42 were simply **divided through by dx** (a procedure scowled upon by the mathematicians) **and the constancy of y imposed** upon the ratios dU/dx, dS/dx, and dV/dx, the result is Eq. A-41. Thus, though the division by dx and imposition of constant y is a procedure unsanctioned by mathematicians, it gives the right answer.

In particular, it is useful for finding expressions for certain partial derivatives, for example, $(\partial U/\partial V)_T$.   By starting with the result

$$dU = T\,dS - p\,dV, \tag{A-43}$$

and "dividing by dV and imposing the constancy of T," one obtains

$$\left(\frac{\partial U}{\partial V}\right)_T = T\left(\frac{\partial S}{\partial V}\right)_T - p, \tag{A-44}$$

which is already proved in Eq. A-41 by the circuitous, but correct, route.

There is really very little more calculus demanded in all of thermodynamics than that given here.   Not that it cannot be introduced if wanted!* However, it may be wise just to rely on a bare minimum of well-understood calculus with which to derive needed equations and solve particular problems. Clever tricks as shortcuts abound in thermodynamics, and some people get great pleasure from the search for them.   Their use, however, helps make the subject mysterious and awe inspiring, and can contribute to the feeling that "I could never think of something as clever as that."

Example A-1. Suppose the internal energy for a certain gas is approximated by

$$U(T, V, n) = \tfrac{5}{2}nRT + n^2 a\left(\frac{1}{V} - \frac{1}{V_0}\right), \tag{A-45}$$

where V_0 is a fixed reference volume and a is a constant.   The equation of state for the gas is approximated by

$$p(T, V, n) = \frac{nRT}{V}\left[1 + \left(b - \frac{a}{RT}\right)\frac{n}{V}\right], \tag{A-46}$$

---

* See, for example, H. Margenau and G. M. Murphy, *The Mathematics of Physics and Chemistry*, 2d ed., D. Van Nostrand Co., Inc., 1956, pp. 15–24, who present a systematic technique for finding some 10,000,000,000 different derivatives in thermodynamics!

where b is a constant and a is the same constant as in Eq. A-45.   Find
(a) $(\partial U/\partial T)_{V,n}$ and (b) $(\partial U/\partial T)_{p,n}$.

Solution: Part (a) is easy.   Since Eq. A-45 already presents $U = U(T, V, n)$,
the differentiation may be done directly with respect to T, treating V and n
as constants just like the a, R, and V_0:

$$\left(\frac{\partial U}{\partial T}\right)_{V,n} = \tfrac{5}{2}nR. \tag{A-47}$$

Part (b) is not easy, since here a different variable is held constant.   The
natural thing to do is to apply Eq. A-33:

$$\left(\frac{\partial U}{\partial T}\right)_{p,n} = \left(\frac{\partial U}{\partial T}\right)_{V,n} + \left(\frac{\partial U}{\partial V}\right)_{T,n}\left(\frac{\partial V}{\partial T}\right)_{p,n}. \tag{A-48}$$

The first term on the right-hand side has been found in Eq. A-47.   The first
term in the product in Eq. A-48 is found simply from differentiating Eq. A-45
with respect to V, holding T and n constant:

$$\left(\frac{\partial U}{\partial V}\right)_{T,n} = -\frac{n^2 a}{V^2}. \tag{A-49}$$

It is not immediately obvious how to find $(\partial V/\partial T)_{p,n}$ from Eq. A-46, since
Eq. A-46 is easy to solve for $p = p(T, V, n)$, but not easy to solve for $V = V(p, T, n)$.   We can take advantage of what is easy to find by using Eq. A-22:

$$\left(\frac{\partial V}{\partial T}\right)_{p,n} = -\frac{(\partial p/\partial T)_{V,n}}{(\partial p/\partial V)_{T,n}}. \tag{A-50}$$

These derivatives may be found directly from Eq. A-46:

$$\left(\frac{\partial p}{\partial T}\right)_{V,n} = \frac{nR}{V}\left[1 + \left(b - \frac{a}{RT}\right)\frac{n}{V}\right] + \left(\frac{nRT}{V}\right)\left(\frac{a}{RT^2}\right)\left(\frac{n}{V}\right) \tag{A-51}$$

or

$$\left(\frac{\partial p}{\partial T}\right)_{V,n} = \frac{nR}{V} + \frac{n^2 Rb}{V^2}. \tag{A-52}$$

Similarly (the reader should check the calculus),

$$\left(\frac{\partial p}{\partial V}\right)_{T,n} = -\frac{nRT}{V^2}. \tag{A-53}$$

Therefore, Eq. A-50 yields

$$\left(\frac{\partial V}{\partial T}\right)_{p,n} = \frac{V + nb}{T}. \tag{A-54}$$

The result has been obtained as a function of V, T, and n, though one would normally think of $(\partial V/\partial T)_{p,n}$ as explicitly a function of p, T, and n. If one needed to replace the V in Eq. A-54 by its value from Eq. A-46, the quadratic equation, Eq. A-46, would have to be solved for V.

We can now substitute Eqs. A-47, A-49, and A-54 into Eq. A-48 to obtain

$$\left(\frac{\partial U}{\partial T}\right)_{p,n} = \frac{5}{2} nR - \frac{n^2 a(V + nb)}{V^2 T}. \tag{A-55}$$

It was a lot of work, but it might possibly have been worth it!

EULER'S THEOREM

Euler's theorem treats functions Φ of x and y (and any number of other variables), which are called *homogeneous* and *nth order* in their variables, that is, ones for which

$$\Phi(gx, gy) = g^n \Phi(x, y), \tag{A-56}$$

where g is any arbitrary number. If each variable is made g times as large, the function Φ becomes g^n times as large. For large enough systems, the volume, energy, and entropy are homogeneous and of first order in the n_i at fixed temperature and pressure. If each mole number is made g times as large, V, U, and S all become g times larger. It is precisely this feature that distinguishes what we call "large systems."

If Eq. A-56 is differentiated with respect to g, using the chain rule on the left-hand side,

$$\frac{\partial \Phi(gx, gy)}{\partial(gx)} \frac{d(gx)}{dg} + \frac{\partial \Phi(gx, gy)}{\partial(gy)} \frac{d(gy)}{dg} = ng^{n-1} \Phi(x, y), \tag{A-57}$$

$$\frac{\partial \Phi(gx, gy)}{\partial(gx)} x + \frac{\partial \Phi(gx, gy)}{\partial(gy)} y = ng^{n-1} \Phi(x, y). \tag{A-58}$$

This result holds for all values of g; thus g can be replaced by any value one chooses. If we choose $g = 1$, Eq. A-58 becomes

$$n\Phi(x, y) = x \frac{\partial \Phi(x, y)}{\partial x} + y \frac{\partial \Phi(x, y)}{\partial y}. \tag{A-59}$$

and this result, where Φ has been brought to the left side, is **Euler's theorem** for homogeneous, nth-order functions. It relates the value of the function to the values of its derivatives. The $\Phi(x, y)$ on the left-hand side is nth order in x and y. The x and y on the right side are, of course, first order, and the derivatives are of one less order than is Φ.

An example perhaps clarifies this result.   Let

$$\Phi = ax^n + by^n + cx^{n-m}y^m. \tag{A-60}$$

This function is clearly nth order in x and y.   Its derivatives are:

$$\frac{\partial \Phi}{\partial x} = nax^{n-1} + (n-m)cx^{n-m-1}y^m, \tag{A-61}$$

$$\frac{\partial \Phi}{\partial y} = nby^{n-1} + mcx^{n-m}y^{m-1}. \tag{A-62}$$

The right-hand side of Eq. A-59 is, then,

$$nax^n + (n-m)cx^{n-m}y^m + nby^n + mcx^{n-m}y^m, \tag{A-63}$$

which is just n times Eq. A-60, as demanded by Euler's theorem, Eq. A-59.

For the homogeneous *first*-order functions considered in Sec. 12, n is, of course, unity:

$$\Phi(x, y) = x\frac{\partial \Phi(x, y)}{\partial x} + y\frac{\partial \Phi(x, y)}{\partial y} \qquad \text{(homogeneous, first order).} \tag{A-64}$$

Thus we may note in, say, Eqs. 12-4 that V, U, and S are homogeneous first-order functions of the n_i at fixed T and p.   Therefore, using Eq. A-64, we may immediately write

$$V = \sum_k n_k \left(\frac{\partial V}{\partial n_k}\right)_{T, p, n_{j \neq k}}, \tag{A-65}$$

with similar expressions for U and S.   From the definitions, Eqs. 12-8, of the partial molar quantities v_i, u_i, and s_i, the results of Eqs. 12-13 to 12-15 are obtained directly:

$$V = \sum_k n_k v_k, \qquad U = \sum_k n_k u_k, \qquad S = \sum_k n_k s_k. \tag{A-66}$$

The use of Euler's theorem to obtain these and similar results is more satisfying to many people than the methods employed in Sec. 12.

APPENDIX B: LEGENDRE TRANSFORMS

The natural variables for the internal energy as a thermodynamic potential are S, V, and n_i, as shown by the Gibbs equation:

$$dU = T\, dS - p\, dV + \sum_k \mu_k dn_k. \tag{B-1}$$

From this, if one knew

$$U = U(S, V, n_i), \tag{B-2}$$

one could find T, p, and μ_k simply by differentiating U:

$$T = \left(\frac{\partial U}{\partial S}\right)_{V, n_i}, \qquad -p = \left(\frac{\partial U}{\partial V}\right)_{S, n_i}, \qquad \mu_k = \left(\frac{\partial U}{\partial n_k}\right)_{S, V, n_{j \neq k}}. \tag{B-3}$$

Suppose, however, one knew

$$U = U(T, V, n_i) \tag{B-4}$$

and not Eq. B-2. How much information is this, in comparison with Eq. B-2? It is clear that Eq. B-4 could be obtained from knowing Eq. B-2 by eliminating S between Eq. B-2 and the first of Eqs. B-3. The question is whether it is possible to go the other way and find Eq. B-2 from Eq. B-4.

By differentiation of Eq. B-4 one could obtain the derivatives present in the expression for the total differential of U:

$$dU = \left(\frac{\partial U}{\partial T}\right)_{V, n_i} dT + \left(\frac{\partial U}{\partial V}\right)_{T, n_i} dV + \sum_k \left(\frac{\partial U}{\partial n_k}\right)_{T, V, n_{j \neq k}} dn_k. \tag{B-5}$$

But these derivatives are not related so simply to the important state functions as are Eqs. B-3. What these derivatives represent may be found from Eq. B-1 by the method discussed in Eqs. A-39 through A-44. " Dividing Eq. B-1 by dT and imposing the constancy of V and n_i" yields

$$\left(\frac{\partial U}{\partial T}\right)_{V, n_i} = T\left(\frac{\partial S}{\partial T}\right)_{V, n_i}. \tag{B-6}$$

Similarly for the other derivatives present in Eq. B-5:

$$\left(\frac{\partial U}{\partial V}\right)_{T, n_i} = T\left(\frac{\partial S}{\partial V}\right)_{T, n_i} - p, \tag{B-7}$$

$$\left(\frac{\partial U}{\partial n_k}\right)_{T, V, n_{j \neq k}} = T\left(\frac{\partial S}{\partial n_k}\right)_{T, V, n_{j \neq k}} + \mu_k . \tag{B-8}$$

Thus, knowledge of Eq. B-4 would permit immediate determination of Eqs. B-6, B-7, and B-8 simply by differentiation. But it is impossible to find S, p, or μ_k from Eqs. B-6 to B-8 without performing an integration, and the value of the integral is undetermined to within an arbitrary integration constant. Thus S, p, and μ_k cannot be determined from knowing simply $U(T, V, n_i)$. It is true that finding $U(T, V, n_i)$ has the advantage of being easier experimentally than finding $U(S, V, n_i)$, since temperature is easier to control than entropy (and since the function of temperature does contain less information). But it has the disadvantage of not being good for much, once it is found!

A specific example will perhaps clarify the loss of information in passing to Eq. B-4 from B-2. An approximate mathematical expression of the fundamental relation for a very dilute monatomic gas (like He, Ne, A, Kr, Xe) is

$$U = U(S, V, n) = an^{5/3}V^{-2/3}e^{2S/3nR} \tag{B-9}$$

where a is a constant that can be determined for the particular gas (see F. C. Andrews, *Equilibrium Statistical Mechanics*, John Wiley & Sons, New York, Section 20). (See also Prob. 13-4.) For fixed n and V, a plot of U versus S resembles Fig. B-1. The slope of this curve is shown by the first of

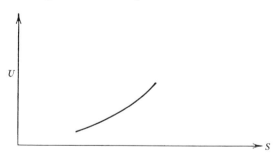

Fig. B-1. Plot of U versus S for monatomic dilute (ideal) gas, V and n fixed.

Eqs. B-3 to be the temperature. From Eq. B-9, the reader can immediately verify (see Probs. 13-5 and 13-6) the thermal equation of state and also the so-called caloric equation of state:

$$p = p(T, V, n) = \frac{nRT}{V} \tag{B-10}$$

$$U = U(T, V, n) = \tfrac{3}{2}nRT.$$ (B-11)

Now, suppose one knew Eq. B-11 instead of Eq. B-9.   One could try to find $U(S, V, n)$ by replacing T in Eq. B-11 by $(\partial U/\partial S)_{V,n}$ and solving the resulting differential equation:

$$U = \frac{3}{2}nR\left(\frac{\partial U}{\partial S}\right)_{V,n},$$ (B-12)

$$dS = \frac{3}{2}nR\frac{dU}{U} \qquad (V \text{ and } n \text{ constant}),$$ (B-13)

$$S = \frac{3}{2}nR \ln U + C \qquad (V \text{ and } n \text{ constant}),$$ (B-14)

$$U = \varphi(V, n)e^{2S/3nR},$$ (B-15)

where φ is an arbitrary function of V and n.   The last step followed from Eq. B-14 by writing it as $2S/3nR = \ln U$ plus the logarithm of a different constant, then taking the exponential of both sides of the equation.   The integration constant can be a completely arbitrary function of V and n, since V and n are constants in the equation.   The single curve of Fig. B-1 could here be any of the whole family of curves shown in Fig. B-2.

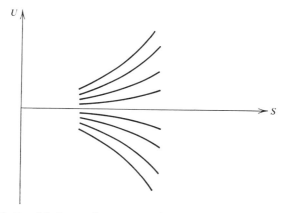

Fig. B-2. Possible forms of U versus S given $U(T, V)$ rather than $U(S, V)$.

The loss of information in Eq. B-15 compared with Eq. B-9 is clear. There is nothing arbitrary about Eq. B-9, in which U is a function of its appropriate variables.   Yet when a derivative of U is used as an independent variable, the result is a differential equation, and there is a whole family of solutions of this, due to the integration constant.   It is impossible to find the temperature from Eq. B-15 by differentiating with respect to S; there is

no unique slope to the curves in Fig. B-2.   However, from Eq. B-7 the slope
is unique and finding T or p is easy.

In order to retain the complete information contained in $U = U(S, V, n_i)$,
what is needed is a way to represent the unique curve of, say, Fig. B-1 in
terms of the slope of that curve (or derivative of the equation).   The line
shown in Fig. B-1 is the *locus of all points* given by the particular relation
between U and S.   However, a line may just as well be viewed as the *envelope
of the family of tangent lines*, as shown in Fig. B-3.   Whereas each point in

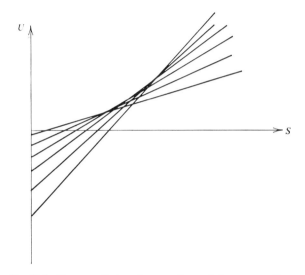

Fig. B-3. U versus S viewed as envelope of the tangent lines.

Fig. B-1 is specified by two numbers, U and S, each straight line in Fig. B-3
is specified by two numbers, the slope, $T = (\partial U / \partial S)_V$, and the intercept on
the U axis, which we shall call A.   Just as a relation $U = U(S)$ selects a
subset of points corresponding to the line of Fig. B-1, a relation $A = A(T)$
selects a subset of all possible tangent lines.   When this subset corresponds
exactly to the same curve as $U(S)$, then either picture is completely equivalent;
both are fundamental representations for the system.

The equation of the line tangent to the curve at a given point is found
immediately from consideration of Fig. B-4.   Given that the straight line
passes through both the point (S, U) and the point (O, A), its equation is

$$\left(\frac{\partial U}{\partial S}\right)_V = T = \frac{U - A}{S} \tag{B-16}$$

or

$$A = U - TS. \tag{B-17}$$

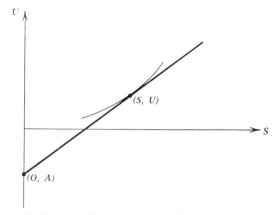

Fig. B-4. Finding the equation of the tangent line.

A quantity like A in which an independent variable is replaced by one of the derivatives of a function is called a **Legendre transform** of that function.

To summarize: We are given a fundamental relation in terms of one set of variables: (1) $U = U(S, V, n_i)$. We desire to use (2) $T = (\partial U/\partial S)_{V, n_i}$ as an independent variable in place of S. The procedure is to find the Legendre transform of the function U, (3) $A = U - TS$ and eliminate the unwanted quantities U and S from these three equations by substitution. This leaves the new fundamental representation

$$A = A(T, V, n_i) \tag{B-18}$$

in the desired variables. The total differential of A is

$$dA = -S\, dT - p\, dV + \sum_k \mu_k\, dn_k, \tag{B-19}$$

which shows how S, p, and μ_i may be calculated from A:

$$S = -\left(\frac{\partial A}{\partial T}\right)_{V, n_i}, \qquad p = -\left(\frac{\partial A}{\partial V}\right)_{T, n_i}, \qquad \mu_k = \left(\frac{\partial A}{\partial n_k}\right)_{T, V, n_{j \neq k}}. \tag{B-20}$$

In the example of the dilute monatomic gas, $A(T, V, n)$ is found by eliminating U and S from the following: Eq. B-9, the equation $T = (\partial U/\partial S)_{V, n}$, and Eq. B-17 for A. The result is

$$A = \frac{3}{2} nRT \ln\left(\frac{2aen^{2/3}}{3RTV^{2/3}}\right) \tag{B-21}$$

where e is the base of the natural logarithms. The reader is asked to check this equation in Prob. 13-7. He also can verify easily that Eq. B-21 generates the same results for Eqs. B-10 and B-11 as did Eq. B-9 (see Probs. 13-8 and 13-9).

Once we have the Legendre transforms, we become aware that they have a host of uses beyond simply being fundamental representations of the system. Tabulations of certain functions permit immediate calculation of the heats of various processes (at given temperatures and fixed pressure). Tabulations of others permit calculation of equilibrium constants. Tables of still others permit correcting those equilibrium constants as the temperature is changed. Other uses are the variety of Maxwell relations stemming from the various Gibbs equations. For example, what is the quickest and most obvious way of deriving the important Eq. 13-34,

$$\left(\frac{\partial S}{\partial V}\right)_{T,\, n_i} = \left(\frac{\partial p}{\partial T}\right)_{V,\, n_i}, \tag{B-22}$$

from Eq. B-1? If one plays around with Eq. B-1 for a while, one very likely convinces oneself that the best way is to define the new function $A = U - TS$ and use the Maxwell relation that results.

APPENDIX C: UNIT CONVERSIONS AND VALUES OF CONSTANTS

Energy	erg = dyne-cm = g-cm^2/sec^2 joule = 10^7 ergs = watt-sec thermochemical calorie = 4.1840×10^7 ergs
Gas constant	$R = 1.98726$ cal/deg-mole $R = 0.08205967$ l-atm/deg-mole $R = 8.31469$ joules/deg-mole
Electricity	coulomb = amp-sec joule = volt-coul volt = ohm-amp $\mathscr{F} = 96{,}493$ coul/equiv = 23,062.4 cal/volt-equiv
Pressure	standard atmosphere = 1,013,250 dynes/cm^2 atm = 760 mm Hg at 0°C and g^0 gravity atm = 760 torrs
Gravity	g^0 = 980.665 cm/sec^2
Length	1 in = 2.54000 cm
Avogadro's number	$N_0 = 6.02308 \times 10^{23}$ mole^{-1}
Standard molar volume	1 mole of ideal gas at 1 atm and 273.15°K occupies 22,414.6 cm^3

APPENDIX D: REFERENCES

The following list of books has been pared to a very few which the author considers of special interest to readers of this book. No one book can make its points uniformly clear to any reader. Students should read widely in an effort to understand the principles and applications of thermodynamics.

Andrews, F. C., *Equilibrium Statistical Mechanics*, John Wiley & Sons, New York, 1963, second edition in preparation. The introductory statistical mechanics text that the author finds most comprehensible.

Denbigh, K., *The Principles of Chemical Equilibrium*, Cambridge University Press, London, 1955, available in paperback edition, 1964. A thorough, advanced text, stressing chemical equilibria and real solutions.

Glasstone, S., *Thermodynamics for Chemists*, D. Van Nostrand Co., New York, 1947. Somewhat dated, and copies are hard to find. A thorough advanced text, carefully covering thermodynamics. Contains many extremely clear explanations, many ideas not discussed elsewhere, and an excellent assortment of problems.

Hill, T. L., *Thermodynamics for Chemists and Biologists*, Addison-Wesley Publishing Co., Reading, Massachusetts, 1968. Coverage of special advanced topics: surfaces, adsorption, elasticity, membranes, electric fields, small systems, and irreversible processes.

Lewis, G. N., **Randall,** M., **Pitzer,** K. S., **and Brewer,** L., *Thermodynamics*, 2d ed., McGraw-Hill Book Co., New York, 1961. A thorough advanced text stressing experimental applications and numerical data.

Pimentel, G. C., **and Spratley,** R. D., *Understanding Chemical Thermodynamics*, Holden-Day, San Francisco, 1969, available in paperback. An intuitive, relaxed little book on understanding the thermodynamics of chemical reactions.

Prausnitz, J. M., *Molecular Thermodynamics of Fluid-Phase Equilibria*, Prentice-Hall, Englewood Cliffs, N.J., 1969. An outstanding presentation of the experimental and theoretical complexities of multicomponent problems of the type discussed in Secs. 26–29 of this book.

Van Ness, H. C., *Understanding Thermodynamics*, McGraw-Hill Book Co., New York, 1969, available in paperback. An appealing, fresh little book, stressing intuitive understanding of basic concepts in thermodynamics.

APPENDIX E: HINTS AND ANSWERS
TO PROBLEMS

5-1. $-p_0(V_2 - V_1)$; $p_0(V_2 - V_1)$.

5-2. 105 l-atm.

5-3. *Hint:* Relate W to area of ellipse. $W = 480\pi$ l-atm.

6-1. Ha!

6-2. What is the meaning of the word "energy"?

6-3. How many experiments have men performed on the universe, as distinguished from tiny pieces of it?

6-4. What is the difference between scientific belief and faith?

7-3. The heat interaction with the lake is masked by the large size of the lake, thus its temperature does not noticeably change.

7-5. *Hint:* See Eqs. 11-34 and 11-35. Efficiency $= 1 - |Q_2|/|Q_1|$.

10-1. What does the sentence mean? In the reversible heating of some water, is the water at $15°$ in equilibrium with the water at $16°$? What minor change would make the statement correct?

10-2. For reversible processes the statement is completely wrong. Why? For adiabatic and zero-work processes it is right. Why?

10-3. Quasistatic is a necessary condition for reversibility, but is it sufficient?

11-1. The second law merely says that if Humpty is put together again, there must be some permanent changes in the King's horses and/or the King's men. A related question, whether the second law means you cannot unscramble an egg, has been answered, "just feed it to a chicken," and "you can maybe do it, but not without leaving a mess in the kitchen."

11-2. 0.497.

11-3. $295/26 = 11.3$.   Why, then, are not heat pumps always used instead of furnaces?

11-12. 0.367 kcal.

11-13. 17×10^{10} kcal/day; 17 m.   In principle, how might the thermal pollution problem be beaten by (a) using heat engines, and (b) using some other means?   *Hint:* See Prob. 30-22.   *Note:* Nuclear power plants seem to be going in the wrong direction, since their boiler temperatures are only 290°C.

17-1. 0.416 atm.

18-1. 0.444 atm for O_2, 1.77 atm for N_2.

18-2. $p_r = 11.9$, $T_r = 2.17$.   The experimental value of v is 0.071 l/mole.

18-4. $W = nRT \ln (p_2/p_1)$.

18-5. -4280 cal.

18-6. *Hint:* One mole of material has volume $v = x_1 v_1 + x_2 v_2$.   Also, $x_1 + x_2 = 1$.

18-9. $b = 0.0562$ l/mole, $a = 6.50$ l²-atm/mole².

18-10. 113 atm.

18-11. $v_c = 0.124$ l/mole experimentally, $v_c = 3b = 1.69$ l/mole using van der Waals' equation.

18-14. $W_1 = -2450$ cal, $W_2 = -19{,}240$ cal, $W_3 = -666$ cal.

18-15. $W = -nRT \ln (V_2/V_1) + n^2 RT(b - a/RT)(V_2^{-1} - V_1^{-1})$.

18-16. $\alpha = 1/T$, $\beta = 1/p$.

18-17. $W = nRT \ln (nRT/V_2 \, p_1) = -4290$ cal.

18-19. 18.9 l-atm or 458 cal.

19-3. 4.8 cal/deg-mole.

19-4. Zero

19-6. -44.1 kcal, -100 kcal.

19-7. 428 m.   Increased.

19-8. 21.26 kcal.

19-9. 5.69 cal/deg-mole, 1.044.

20-1. $p_1 V_1/T_1 = p_2 V_2/T_2$.

20-3. $C_V(T, V_1) = C_V(T, V_0) + 2n^2 a(V_0 - V_1)/T^2 V_0 V_1$.

21-3. 31.1 eu.

21-4. 8.94 eu.

21-5. -0.316 eu.

21-6. $\Delta S = C_p \ln (T_2/T_1) - nR \ln (p_2/p_1)$.

21-7. (a) 3.24 eu; (b) 18.1 eu; (c) 0.351 eu.

21-8. $\Delta S = -2050$ eu, $\Delta H = -719.4$ kcal.

21-9. Final temperature is $60°C$. $\Delta S = 0.445$ eu. *Note:* Slide rule accuracy is inadequate for this problem.

21-11. (a) 180 cal/deg, -180 cal/deg, 0; (b) 180 cal/deg; for the reservoir $\Delta S = \Delta H/T = -(60{,}000 \text{ cal})/360 \text{ deg} = -166.7$ eu; for the composite $\Delta S = 13.3$ eu. Note the increase in total entropy of the isolated composite during this irreversible process.

21-12. Resistor is in steady state, its properties do not change, so $\Delta S = 0$. For the water, $\Delta S = 7.07$ joules/deg.

21-13. The surroundings have $\Delta S = 0$, since they have a work-only inter-action which, from their viewpoint, could be reversible. For the resistor $\Delta S = C \ln (T_2/T_1)$, and T_2 is found from $\Delta U = C(T_2 - T_1) = i^2 rt$. $\Delta S = 0.99$ eu.

21-15. 7410 cal/mole; the experimental value is 7353 cal/mole.

21-16. (a) $(C_1 T_1 + C_2 T_2)/(C_1 + C_2)$. *Hint to second part:* The total process is reversible and adiabatic, thus $\Delta S = 0$. (b) The final temperature $T = (T_1)^{C_1/(C_1+C_2)}(T_2)^{C_2/(C_1+C_2)}$ is found from equating the sum of the individual entropy changes to zero. (c) Since the overall process is work-only with respect to surroundings, $W = \Delta U$, and ΔU is that found by cooling an object of heat capacity $C_1 + C_2$ from the tempera-ture of (a) to that of (b); $W = (C_1 + C_2)[(C_1 T_1 + C_2 T_2)/(C_1 + C_2) - T_1{}^{C_1/(C_1+C_2)}T_2{}^{C_2/(C_1+C_2)}]$.

21-17. Here the final temperature is T_2. Again $\Delta S = 0$; it is the sum of $C_1 \ln (T_2/T_1)$ for object 1 and $-[C_1(T_2 - T_1) + W]/T_2$ for object 2. The quantity in brackets is the energy lost by 2 (a) to heat up object 1 and (b) to perform the work. The result, $W = C_1 T_2 \ln (T_2/T_1) - C_1(T_2 - T_1)$, is valid for both $T_2 < T_1$ and $T_2 > T_1$.

21-18. $\Delta S_{\text{fus}, v} = \Delta S_{\text{fus}, p} + (\alpha/\beta)(v_s - v_1)$.

21-19. $Q = na(T_2 - T_1) + \frac{1}{2}nb(T_2{}^2 - T_1{}^2) + \frac{1}{3}nc(T_2{}^3 - T_1{}^3)$;
$\Delta S = na \ln (T_2/T_1) + nb(T_2 - T_1) + \frac{1}{2}nc(T_2{}^2 - T_1{}^2)$.

21-20. Since the wall properties are all time independent, the entropy of the wall (insofar as entropy is meaningful for objects not at equilibrium) must remain constant. $\Delta S_{\text{total}} = -200$ cal/373 deg-min $+ 200$ cal/293 deg-min $= 0.146$ eu/min for the two reservoirs. A commonly used description of this is to say that the irreversible processes in the wall generate 0.146 eu/min, and in order for the wall to have $\Delta S = 0$ over time, **negentropy** (that is, negative entropy) in amount 0.146 eu must flow into the wall each minute. No net energy flows into it, but the energy entering is at a higher temperature than the energy leaving, which represents an influx of negentropy. In this sense, the sun provides negentropy for the biosphere, and it is this which drives life.

21-21. *Hint:* Take water at $-10°C$ to ice at $-10°C$ via water at $0°C$ and ice at $0°C$. *Ans:* 0.0345 eu.

22-3. $T_2 = 0.63 \, T_1$.

22-4. $p_2 = 0.316 \, p_1$.

22-5. $T_2 = 0.76 \, T_1$.

22-6. $-21°C$.

22-7. An adiabatic process has $W = \Delta U$. For an ideal gas, $\Delta U = C_V(T_2 - T_1)$. The T_2 may be related to T_1 by Eq. 22-12, and one finds $W = C_V T_1[(V_1/V_2)^{\gamma - 1} - 1]$.

22-8. -1212 cal, -727 cal.

22-9. The integrand $\gamma(\partial p/\partial V)_T$ would have to be expressed explicitly as a function of V and S so that S could be held constant in the V integration. Almost never is explicit S dependence known. Thus equations like Eq. 22-4 are needed.

22-10. (b) $c^2 = (\gamma/M)[RT(v/v - b)^2 - 2a/v]$.

22-11. $C_p(p_1) \ln (T_2/T_1) - nR \ln (p_2/p_1) + na(p_2 - p_1)/RT_2{}^2 = 0$.

22-12. One can derive either $\beta_s = \beta - TV\alpha^2/C_p$ or $\beta_s = \beta/\gamma$. Then substitution yields $\beta_s = 1.28 \times 10^{-12}$ cm^2/dyne.

22-13. One can derive $(\partial T/\partial p)_s = TV\alpha/C_p$. Integration at essentially constant T gives $\Delta T = -0.09°C$.

22-14. "Divide through Eq. 21-5 by dV" and impose the constancy of S. Since $(\partial p/\partial T)_V$ is positive, isentropic expansions always decrease the temperature and isentropic compressions increase it. This effect is important in the operation of the Diesel engine, in which air is rapidly compressed in the cylinder and thus heated almost isentropically. Fuel is injected at nearly the time of full compression, and the compressed air is so hot it ignites the fuel. Thus no spark needs to be provided. The same effect contributes to preignition in ordinary gasoline engines.

23-1. $Q = 9440$ cal, $W = -900$ cal, $\Delta U = 8540$ cal.

23-2. Because this is an ideal gas at constant T, $\Delta U = 0$ and $Q = -W$. Real gases could not be solved so easily. $W = -157$ cal, $Q = 157$ cal.

23-3. 0.112 eu.

23-4. $C_V(V_1)(T_2 - T_1) = -n^2 a(V_2 - V_1)/V_1 V_2$.

23-5. $\mu_J = -n^2 a/C_V V^2$.

23-6. $C_V(V_1) \ln (T_2/T_1) + nR \ln (V_2/V_1) + n^2 Rb(V_2 - V_1)/V_1 V_2$, where T_2 is given in Prob. 23-4 in terms of T_1, V_1, and V_2.

23-7. Do calculations separately for the two steps. The first is an isothermal ideal gas compression to 50 l, having $W = -Q = 34.5$ l-atm. The second is the condensation of $\frac{4}{5}$ of the water present at constant pressure with $W = 40.0$ l-atm and $Q = -497$ l-atm. Totals are $W = 74.5$ l-atm, $Q = -532$ l-atm.

23-8. The difficulty is determining what is the system. Choose 1 mole of gas initially outside that ends up inside. It is squeezed in through the leak under a constant 1 atm of pressure. This increases its energy. $T_{\text{final}} = \gamma T$. This effect is not readily observed because of the small heat capacity of air.

24-2. (a) 6320 cal; (b) -1185 cal; (c) 6320 cal; (d) 5135 cal; (e) 11.23 eu.

24-3. $\Delta H_{\text{total}} = 0$, final $T = 38.2°C$.

24-4. 0.1515 eu.

24-5. 2.3 cm.

24-6. $S = 0.30$ eu/mole, $H - H(0°K) = 3.34$ cal/mole.

24-7. 677 cal/g.

24-8. $\Delta H = -87.0$ cal, $\Delta U = -81.2$ cal.

24-9. 10 l-atm.

24-10. 40.5°C.

24-11. $C_p(p_1)(T_2 - T_1) = (nb - 2na/RT_2)(p_1 - p_2)$.

24-12. $2a/RB$.

24-14. Use Eq. 24-15 in the form $C_p \, \Delta T = (T\alpha - 1)V \, \Delta p$ to get $\Delta T = 0.041$ deg/atm.

24-15. 405°K.

24-18. 0.113.

25-1. (a) -763 cal; (b) 7958 cal; (c) 7958 cal; (d) 7206 cal; (e) 0, (f) 20.7 eu.

25-3. 1.25 mm.

25-4. *Hint:* Presume that the air is saturated with vapor in the bubbling process. Figure the volumes of gases using $pV = nRT$. The presence of CCl_4 increases the total gas volume from the original 10 l by an amount which depends on the vapor pressure, thus a little algebra is required. *Ans:* 0.120 atm.

25-5. (a) 4670 cal/mole; (b) 741 mm.

25-6. (a) $-38.81°C$; (b) $-16°C$. The observed melting point under 3540 atm is $-19.9°C$.

25-7. 6.6 atm.

25-8. 93°C.

25-9. $W = -2.6$ kcal, $Q = 18.9$ kcal, $\Delta S = 50.7$ eu, $\Delta U = 16.3$ kcal, $\Delta H = 17.5$ kcal, $\Delta A = -2.6$ kcal, $\Delta G = -1.4$ kcal.

25-11. (a) $-nRT \ln [(V_2 - nb)/(V_1 - nb)]$; (b) $nRT[V_2/(V_2 - nb) - V_1/(V_1 - nb)] + \Delta A$ [given in part (a)]; (c) $nR \ln [(V_2 - nb)/(V_1 - nb)]$; (d) 0 (this must be proved!); (e) same as (a); (f) negative of (e).

25-12. Thanks to the fact that the gas has zero Joule coefficient, the process would still be isothermal. Only difference is $Q = W = 0$.

25-13. (a) 195°K; (b) 7500 cal, 6130 cal; (c) 1370 cal.

25-15. $W = 0$, $Q = 0$, $\Delta H = 0$, $\Delta S = 0.094$ eu, $\Delta A = 26$ cal, $\Delta G = 26$ cal.

25-16. The two terms in Eq. 25-34 contribute -90 and -54 cal/mole, respectively; total is -144 cal/mole. Compared to $(80)(18) = 1440$ cal/mole at Δh at 0°C, it represents a 10% correction for just a 10°C temperature change.

25-17. 14°C.

25-19. Prove $d\mu$ (T const) $= (1/\rho)(\partial p/\partial \rho)\,d\rho$.   Then find $d\mu$ from the virial series.   Then integrate from ρ_0 to ρ.

26-1. 1.37 l, 0.875, 0.875 atm $= 664$ mm.

26-3. Amagat says $v = \sum_j x_j v'_j$; virial is $v = RT/p + \sum_{jk} x_j x_k B_{jk}$ for mixture and $v'_j = RT/p + B_{jj}$ for pure gas.   Equating implies $\sum_{jk} x_j x_k B_{jk} = \sum_j x_j B_{jj}$ for all values of the mole fractions.   Rewrite this for two components and replace x_2 by $1 - x_1$.   The result must hold for *all* values of x_1, thus it may be differentiated with respect to x_1.   The only way the result of that can hold for all values of x_1 is if the desired relationship holds.

27-1. $x_{\text{benz}} = 0.574$, $y_{\text{benz}} = 0.772$.

27-2. -3.61×10^5 eu.

27-3. $p_{\text{EtOH}} = 18.2$ torrs, $p_{\text{MeOH}} = 52.3$ torrs, $p_{\text{total}} = 70.5$ torrs, $y_{\text{MeOH}} = 0.742$.

27-4. $x_A = 0.744$, $y_A = 0.93$.

27-5. $\Delta U = \Delta H = 0$, $\Delta S = 100.2$ eu $= 4.1$ l-atm/deg, $\Delta A = \Delta G = -29.8$ kcal.

27-6. (*a*) 10.48×10^{-3} l-atm/deg.

27-7. -0.995 eu.

27-8. -52.8 cal/mole.

27-10. 0.0935 mm, doubled.

27-11. $M_2 = g_2 M_1 p/[(p^0 - p)g_1]$.

27-12. 2.23.

28-3. 3.38×10^{-2} mole/l.

28-5. 20.7%.

28-6. 0.5.   This is helpful to the fish.

28-7. $\frac{1}{2}$ mole pure A, $1\frac{1}{2}$ mole azeotrope.   The A comes off the top.

28-8. Multiply Eq. 28-35 by dx_1 and integrate from $x_1 = 0$ to $x_1 = 1$.   Replace x_2 by $1 - x_1$ and integrate each integral by parts.   Note how the integrated terms are zero and one is left with the desired result.

28-9. Set $\ln \gamma_2 = (A/RT)(1 - x_2^2)$.   Use this in Eq. 28-34.   Then $d \ln \gamma_2 = (-2A/RT)(1 - x_2) \, dx_2$, and integration yields $\ln \gamma_1 = Ax_2^2/RT$.

28-10. First prove $(\partial \mu_i/\partial V)_{T, n_j} = v_i (\partial p/\partial V)_{T, n_j} = -(\partial p/\partial n_i)_{T, V, n_{j \neq i}}$ using Eq. A-23. Again go from n_i moles of i in its ideal-gas standard state in three steps: (1) Increase the volume of the ideal gas from $V_0 = (1 \text{ mole})RT/(1 \text{ atm})$ to $\mathscr{V}$, where $\mathscr{V}$ is so large the real mixture is ideal; $\Delta \mu_i = -\int_{V_0}^{\mathscr{V}} RT \, dV/V$. (2) Mix the gases at fixed T and $\mathscr{V}$; $\Delta \mu_i = 0$. (3) Decrease the volume of the mixture from $\mathscr{V}$ to V;

$$\Delta \mu_i = -\int_{\mathscr{V}}^{V} \left(\frac{dp}{dn_i} \right) dV. \quad \text{Then write } \int_{V_0}^{\mathscr{V}} \text{ as } \int_{V_0}^{V} + \int_{V}^{\mathscr{V}}; \text{ replace } \mathscr{V} \text{ by } \infty,$$

perform the trivial integration, and note where needed that the total number of moles must be governed by $y_i = (1 \text{ mole})/n$.

28-11. Poorer.

28-12. 91.5 g.

28-13. (*a*) 30 mm; (*b*) 21 mm; (*c*) 1.43, (*d*) 1.00.

28-14. 1.00 qt.

28-15. $G^E = nwx_1 x_2$, and $\mu_1{}^E = (\partial G^E/\partial n_1)_{T, p, n_2}$. In order to perform the differentiation, G^E must be made explicitly a function of n_1 and n_2, using $x_i = n_i \sum_j x_j$.   Ans: $\mu_1{}^E = RT \ln \gamma_1 = wx_2^2 = w(1 - x_1)^2$. Compare with Eq. 28-25.

28-16. Use Eq. 28-34 to obtain $\ln \gamma_1 = (a + \frac{3}{2}b)x_2^2 - bx_2^3$.

28-18. Use Eq. 12-13 and the first of Eqs. 12-9 to prove $\sum_i n_i \, dv_i = 0$, then divide by n.

28-19. -300 cal, -480 cal.

28-20. See M. L. McGlashan, *J. Chem. Educ.* **40,** 516 (1963).

29-1. 0.513 deg-kg/mole.

29-2. 1.86 deg-kg/mole.

29-3. 26.9 atm.

29.4. 3130 atm.

29-5. $x_{\text{naph}} = 0.297$.   The experimental value is 0.296.

29-6. $\ln (p_{1, A}/p_{1, 0}) = \ln (1 + \delta p/p_{1, 0}) = v_{H_2O}(1 \text{ atm})/RT$. Use $\ln (1 + \varepsilon) = \varepsilon$ for small ε, and get $\delta p/p_{1, 0} = 7.36 \times 10^{-4}$.

29-7. $\Delta p/p_0 = 0.0177$ kg/mole. $\Delta \pi = 24.4$ atm-kg/mole.

29-8. 300 g/mole.

29-9. $2.56 \times 10^{-2}\%$!

29-10. 0.335°C.

29-12. 1.28 kcal/mole.

29-13. $x_{naph} = 0.268$.

29-14. -9.87×10^{-3} deg.

29-15. $\pi = -RT \ln (1 - x_2)/v_1 - wx_2{}^2/v_1$.

29-18. 1.85 times as large, 2.37 times as large.

29-17. *Hint:* The work is $\pi(1000 \text{ l})$, since the scheme of Prob. 29-18 permits in principle the constant-pressure desalination, provided the solution is kept replenished so its concentration does not change.   *Ans:* 0.65 kwhr/1000 l, $6\frac{1}{2}$ cents/1000 l, \$62.70/1000 l.

29-18. Work $= \int \pi \, dV$, and $\pi = \pi^0 V^0/V$ where V^0 is the volume of original sea water and π^0 its osmotic pressure.   This process makes $V^0 - V_f$ of pure water, in contrast to the process of Prob. 29-17, which makes V^0 of pure water for $\pi^0 V^0$ of work.   *Ans:* 1.85 times as large, 2.37 times as large.

29-19. $h_{max} = 3230 \log (1/r)$ m.   At $r = 0.9$, $h = 1.45$ m.

30-1. Molality of $2 = 55.5m_2/m_1 = 55.5x_2/x_1 = 55.5x_2/(1 - x_2)$.   When x_2 is 1% the $(1 - x_2)$ term changes this by 1%, that is, when molality $= 0.56$.

30-2. $\mathscr{E} = \mathscr{E}^0 - (0.0591/n) \log Q$.

30-3. 2.368 eu, -4.737 eu.

30-4. $K_c = 0.500 \text{ l}^2/\text{mole}^2$, $K_x = 1.64 \times 10^{-2}$.

30-5. 43.2 kcal.

30-6. $x_{isob} = 0.714$.

30-7. $\alpha = 0.800$, $K_p = 1.78$ atm.

30-8. $\alpha = 0.0574$.

30-9. 0.341 mole.

30-10. $p = \sum_i p_i = 3.66$ atm.

30-11. 42.7 atm.

30-12. 1%.

30-13. 0.0417 molal.

30-14. 72.0 kcal/mole.

30-15. 278.4 kcal/mole.

30-16. 106.8 kcal/mole.

30-17. $\mu^0_{HCl(g)} + RT \ln p_{HCl(g)} = \mu^0_{H^+} + RT \ln a_{H^+} + \mu^0_{Cl^-} + RT \ln a_{Cl^-}$ or $p_{HCl} = K m^2_{HCl}$.

38-18. $3470°K$.

30-19. 9720 cal/mole.

30-20. -23.06 eu.

30-21. -14.8 kcal/mole.

30-22. $Q = 6.32$ kcal, $W = -109.28$ kcal, as contrasted with $Q = -44.6$ kcal, $W = -71.0$ kcal.

30-23. $\Delta G^0 = 212$ cal, graphite stable. One may change ΔG^0 by 212 cal thus: $\Delta(\Delta G^0) = -212$ cal $= (v_D - v_G) \Delta p$, with $\Delta p = 4630$ atm.

30-24. $K_{sp} = 1.6 \times 10^{-11}$.

30-25. $C = [1 + K/(H^+)][(H^+) - K_w/(H^+)]/[1 + 2K/(H^+)]$, asymptotic form is $(H^+) \rightarrow C$.

30-26. $(H^+) = K_w/(H^+) + C_s - C_s/[1 + K_w/(H^+)K_B]$.

30-27. $2(Fe^{3+}) = -C_1 + C_3 - K + [(C_1 - C_3 + K)^2 + 4K(C_2 + C_3)]^{1/2}$; $(Fe^{++}) = C_2 - C_3 - (Fe^{3+})$; $(I^-) = C_1 - C_3 + (Fe^{3+})$.

INDEX